PELICAN BOOKS

ANIMALS WITHOUT BACKBONES

II

Ralph Buchsbaum was born on 2 January 1907 in Oklahoma, the son of a doctor who was very interested in zoology. Educated at the University of Chicago, where he obtained his Ph.D. in zoology, he held the positions of Assistant, Instructor, and Assistant Professor of Zoology. During the war he was a Captain in the U.S. Air Force and assigned to the Arctic, Desert, and Tropic Information Centre. From 1946 to 1950 he was Research Associate in Zoology in the Institute of Radio Biology and Biophysics at the University of Chicago. He is at present Professor of Zoology, University of Pittsburg, a member of many American scientific associations and also a member of the Marine Biological Association of the United Kingdom. He is married and has two children.

D0551027

ANIMALS WITHOUT BACKBONES

AN INTRODUCTION TO THE INVERTEBRATES

RALPH BUCHSBAUM

VOLUME
TWO

PENGUIN BOOKS

Penguin Books Ltd, Harmondsworth, Middlesex England
Penguin Book Australia Ltd, Ringwood, Victoria, Australia

—

First published in the U.S.A. 1938
Revised edition 1948
Published in Pelican Books 1951
Reprinted 1953, 1955, 1957, 1960, 1961, 1963, 1964, 1966, 1968, 1971
Copyright © Ralph Buchsbaum, 1938, 1948

—

Made and printed in Great Britain
by C. Nicholls & Company Ltd
Collogravure plates by Harrison & Sons
Set in Monotype Times

CONTENTS TO VOLUME TWO

NOTE

DR BUCHSBAUM'S *Animals without Backbones* introduces the
English reader to a new kind of book – a serious scientific work
which is as attractive and easy to read as any natural history. It
differs from most textbooks in its rejection of all detail which
tends to distract interest and attention from main issues, and in
the references it makes to animals that matter to human beings.
It is also outstanding pictorially. There are photographs of
animals alive and at home, as beautiful technically as the
illustrated weeklies have taught us to demand. There are drawings
so sure and vivid that you feel the animal moving in its characteristic way. And there are diagrams so bold and graceful that hard
facts about the structure of animals are easily taken in. Thus
aquarium and laboratory are brought together, and you can
watch a starfish living its starfish life on one page, and can study
its anatomy with dissecting instruments and microscope on
another.

Animals without Backbones is published in two volumes in
the Pelican series. Each volume gives a complete account of
the various groups of animals it deals with, and so can be read
by itself. The two together give a comprehensive survey of the
invertebrate members of the animal kingdom, and if you want
to take pleasure in discovering the serious scientific content of
invertebrate zoology, these are your books.

M. L. JOHNSON

For technical reasons it has been impossible to make the division
of the photographic plates between the two volumes of *Animals
without Backbones* agree with the division of the text into two
volumes. Plates 54–64 of the first volume are concerned with animals
described in the early chapters of this one.

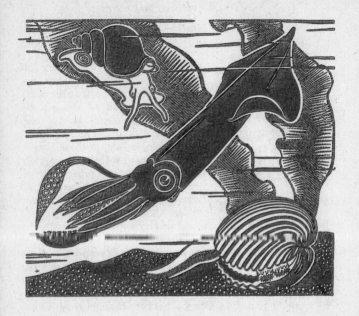

CHAPTER 18

Two Ways of Life – Clam and Squid

THE most generalized, and therefore the most representative, molluscs are the members of the class AMPHINEURA, which includes the chitons described in the preceding chapter. The most specialized molluscs are the members of the groups to which belong the clam and the squid. But before going on to these we shall consider briefly two other classes.

GASTROPODS

THE class GASTROPODA includes such common animals as snails and slugs, limpets and whelks, and is by far the largest of the classes of molluscs.

Most gastropods show all of the chief molluscan features: a protective shell, a mantle, a large fleshy foot, a dorsally placed visceral mass, a radula, and usually one or more gills. In the

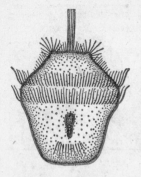

TROCHOPHORE of a gastropod.
(After Patten.)

possession of a well-developed head with eyes and sensory tentacles they are more like our idea of the primitive mollusc than is the chiton. But they are highly modified in the possession of a coiled shell and an asymmetrical organization of the visceral mass. The asymmetry results from atrophy of most of the visceral organs of one side, leading to the coiling of visceral mass and shell. The spirally coiled shell is a very compact arrangement for the disproportionately long visceral mass of gastropods. If the viscera and shell were in the form of a long straight cone, they would be almost unmanageable and a serious impediment to locomotion.

In the chiton the mouth and anus occur at opposite ends of the body. In most grastropods the anus opens anteriorly and lies above the head. The advantages of this arrangement are clear enough in an animal that lives in a shell with only one opening. Of the exact manner in which this has been brought about we are less certain. The explanations that have been given are based upon the development of gastropod larvas.

The molluscan larva starts out as a TROCHOPHORE with an equatorial girdle of cilia. As growth continues, the band of cilia becomes expanded, often into very large ciliated lobes, which serve to propel the larva about and bring food to the mouth. This second free-swimming larva with the expanded ciliated zone is called the VELIGER and is peculiar to molluscs. It occurs in all classes except the one to which belong the squids and octopuses. While the trochophore has the larval organs seen in marine larvas of this type, the veliger is characterized by the development of adult organs, such as the shell and foot.

The veliger is at first bilateral and has an anterior mouth, a posterior anus, a dorsal shell, and a ventral foot, as in the chitons. As development proceeds, the digestive tube is bent downward and forward until it lies near the mouth. This approximation of mouth and anus occurs in some members of all classes of molluscs except the Amphineura. But what follows is peculiar to gastropods. While the head and foot remain stationary, the visceral mass is rotated through an angle of

180°, so that the anus and the mantle cavity that surrounds it are carried upward and finally come to lie dorsal to the head. In addition, the organs on one side of the body fail to develop; and as a result of this unequal growth, the visceral mass and mantle (and the shell secreted by the mantle) become spirally coiled.

In most gastropods it is the organs of the original left side that degenerate. But since the visceral mass is rotated through 180°, the adult appears to lack the kidney, the gill, and one of the chambers of the heart (atrium) on the right side. The nervous system of these gastropods remains bilateral and uncoiled, but becomes twisted into a figure-of-eight when the viscera rotate. In some gastropods there is a reversal

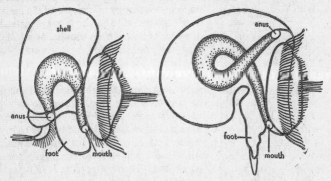

Left, a VELIGER before the rotation of the viscera. *Right*, a veliger after the rotation of the viscera; the anus now lies above the mouth. (Based on Patten and Robert.)

of the rotation of the viscera (see below) and the nerve cords are untwisted.

Only the more primitive gastropods have a free-swimming trochophore. Most marine forms pass through the trochophore stage while still confined within the protective capsule in which the eggs are laid, and emerge as veligers. Some gastropods pass through even the veliger stage within the capsule and emerge as young adults. In land and freshwater gastropods, development of the eggs is modified, and usually there is no recognizable veliger stage. In some cases the eggs are not laid at all, but develop within the body of the parent.

For all its advantages, a shell is a handicap to active locomotion. And there is a tendency among many groups of gastropods toward reduction or even complete loss of the shell, accompanied by an uncoiling and an untwisting of the visceral mass. That

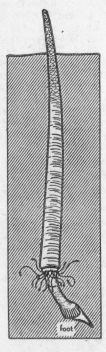

A SCAPHOPOD lies buried in the sand with only the upper end of the shell protruding into the water. (Modified after Sars.)

these forms have descended from typical gastropods is shown in their larval development. The larvas have a coiled shell and undergo twisting, followed by an untwisting and by loss of the shell and uncoiling of the viscera. As is generally true in animal evolution, organs once lost are not regained, although some substitute may be formed eventually from some other region of the body. Thus, gastropods descended from forms which have undergone atrophy of one side of the body have only one gill and one kidney (instead of a pair of each), even though, as adults, they are uncoiled, and have reverted to an external bilateral symmetry.

Gastropods are most successful in the water; but many, like the snails and slugs, have invaded the land. These are called the *pulmonate* gastropods because they have a modified mantle and mantle cavity which acts as a 'lung' for air-breathing. Many pulmonates have gone into fresh water; but, as has been pointed out before, organs once lost do not reappear again, and, though aquatic, these snails have no gills. They must come to the surface periodically to take air into the lung. Some freshwater snails have gills, but these have descended directly from marine snails instead of land types.

TOOTH SHELLS

A SMALL class of molluscs are the 'tooth shells' or SCAPHOPODA, in which the shell is shaped like a miniature elephant's tusk, tubular and open at both ends. There is a mantle, a poorly-developed head bearing a number of extensible filaments that serve as sense organs and aid in capturing prey, a radula, and a muscular burrowing foot. The gills are lost, and the mantle serves as the respiratory organ. A current of water is maintained in and out of the upper end of the shell, as the animal lies

buried almost completely in the sand.
The larva goes through free-swimming
trochophore and veliger stages.

BIVALVES

THE clam, oyster, scallop, and others of
the molluscs with two shells are often
called 'bivalves' (two valves) and comprise the class PELECYPODA. The name of
the group means 'hatchet foot' and
refers to the shape of the foot in many
pelecypods. Most members of the group
are marine, but some clams are very
abundant in fresh waters. The description that follows applies in general to
almost any of our common fresh-water clams.

The SHELL consists of
three layers.

The CLAM is flattened from side to side. The two shells or
valves, which represent right and left sides, are fastened to each
other dorsally by an elastic horny LIGAMENT. The gape of the
shells is ventral. Near one end of the ligament is an elevated
knob, the UMBO. The end of the animal nearer the umbo is the
anterior end. At the opposite or posterior end are the openings
through which currents of water enter and leave the clam.

The umbo represents the oldest part of the SHELL. As the
animal grows, the mantle secretes successive layers of shell, each
projecting beyond the last one laid down. This results in a series
of concentric lines of growth which mark the external surface
of the shell and represent the successive outlines of its ventral
margin.

The shell consists of three layers. The dark, horny, *outer layer* (perio-
stracum) forms the ligament and protects the calcareous shell from
being dissolved by carbonic acid in the water. It is thin and is usually
eroded from the older parts of the shell, such as the umbo. The *middle
layer* (prismatic layer) consists largely of crystals of calcium carbonate
arranged perpendicularly to the surface of the shell. The innermost,
or *pearly layer* (nacreous layer) consists mostly of thin sheets of calcium
carbonate laid down parallel to the surface of the shell. The first two
layers are secreted only by the edge of the mantle, and hence show the
concentric markings of discontinuous growth. The inner, pearly layer
is laid down by the whole surface of the mantle and has a smooth,
lustrous surface.

A *pearl* may be secreted by the mantle as a protection against some foreign body, usually a parasite such as the larval stage of a fluke. The larva enters the mantle and becomes enclosed in a sac formed by the growth of the mantle epithelium, which secretes thin, concentric layers of pearly substance around the foreign body.

When undisturbed, the clam lies partly buried in the sand or mud with the ligament up and the shells slightly agape ventrally. In this position the animal protrudes its fleshy foot and burrows through the mud like an animated ploughshare. First, the pointed foot is extended forward into the mud and anchored by a turning or by a swelling of the free end (due to an influx of blood into a cavity within the foot). Then, as the muscles of the foot contract, they draw the body of the clam forward. Such a slowly moving

FORMATION OF A PEARL. *Left*, a parasite lodges between the shell and the mantle epithelium. *Middle*, it is almost completely enclosed in a sac formed by the epithelium, which secretes thin, concentric layers of pearly substance. *Right*, a pearl of good size has surrounded the parasite and prevented it from harming the clam.

animal, with a shell that is heavy and cumbersome to carry about, could hardly run down its prey. Instead, like so many other sedentary animals, the clam has evolved a method of drawing water through its body and straining out the microscopic organisms and other nourishing organic particles contained in the water. For protection it relies on its heavy shell and retiring habits.

The shells are held agape by the elasticity of the ligament. They are closed by the contraction of two large MUSCLES. Near these are smaller muscles which extend and retract the foot; and attached to the shell along a line close to, and parallel with, its ventral margin is a row of small muscles which retract the edge of the mantle. When the shell is lifted back (after cutting the muscles), its inner surface will show 'scars', which represent the former attachments of all these muscles. Also, it will be seen that the dorsal margin of the shell has long ridges and irregular tooth-like projections, the HINGE TEETH, that fit into grooves or pits in the opposite shell. This interlocking arrangement fits the two valves together.

The visceral mass lies dorsally, most of it between the two large muscles that close the shell. The MANTLE covers the visceral mass

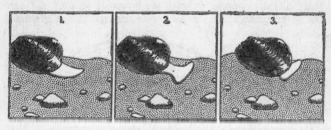

A MOVING CLAM extends its foot into the sand ...

the tip of the foot swells and acts as an anchor ...

the muscles of the foot contract, drawing the body of the clam forward ...

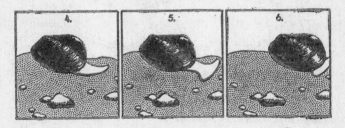

the foot is again extended ... the tip is anchored

... and the body is again

drawn forward.

and extends ventrally as two mantle lobes, one just underneath each shell. The space between the mantle lobes is the MANTLE CAVITY. At the posterior end the lobes are thickened locally and approximated at certain points to form the openings for entrance and exit of water. These extend out just beyond the margins of the shells, when the shells are agape. A current of water passes into the mantle cavity through the ventral or *incurrent opening* and out through the dorsal or *excurrent opening*.

On removing one mantle lobe, the mantle cavity and its organs are exposed. From the mid-ventral region of the visceral mass the foot extends into the mantle cavity and out between the shells. Between the foot and the mantle lobe, on each side, a pair of sheetlike double folds, the GILLS, hang freely into the mantle cavity. The gills have a sievelike structure, being perforated by microscopic PORES, and are covered with cilia. The beating of these cilia draws water through the incurrent opening and into the mantle cavity. The water passes through the microscopic

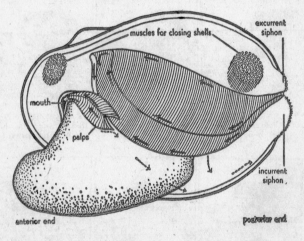

The FEEDING MECHANISM is ciliary; food particles, caught on the surface of the gills, are carried to the mouth, as shown by the *solid arrows*. Rejected particles are removed from gills and palps, as shown by *dotted arrows*.

pores, leaving suspended particles on the surface of the gills. Within the gills the water flows up the WATER TUBES, vertical channels formed by partitions that subdivide the cavity between the inner and outer walls of a gill. The water tubes open dorsally into the DORSAL GILL PASSAGES, which run one above each gill and open posteriorly near the excurrent opening, through which the water leaves the clam.

The rate at which water passes through the mantle cavity has been measured in certain marine clams and oysters. For an animal of

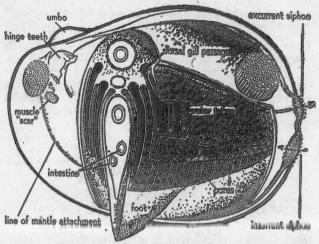

The direction of the RESPIRATORY CURRENTS of the clam is shown by the arrows. A piece of one gill has been cut away to show that the space between the inner and outer walls of the gill is divided into vertical water tubes. The pores of the gill are microscopic and are shown here greatly enlarged. The *anterior part of the clam has been removed.* In the *sectioned surface* the relations of the water tubes to the dorsal gill passages can be seen. The intestine appears more than once in the section because it coils back and forth in the foot.

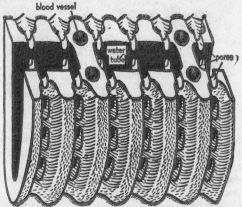

A SMALL PORTION OF A GILL enlarged to show how the partitions, in which the blood vessels run, divide the space between the inner and outer walls of the gill into vertical water tubes which communicate with the mantle cavity through microscopic pores.

average size, the minimum rate averages about 2.5 litres (almost 3 quarts) an hour.

FOOD PARTICLES left on the surfaces of the gills by this steady stream of water are distinguished, mostly by their small size, from silt and other undesirable materials during their passage to the mouth. Heavy particles of sand or mud simply drop from the surface of the gill to the edge of the mantle, are carried backward by cilia on the mantle, and are expelled posteriorly.

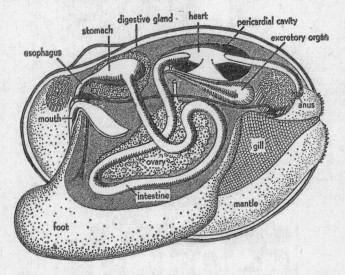

Clam with mantle, one pair of gills, and part of foot cut away to show the DIGESTIVE SYSTEM and other organs.

The lighter particles become entangled in mucus secreted by the gills and are carried, always by beating cilia, to the ventral edge of the gill and then forward until they meet the ciliary tracts on the PALPS, a pair of folds on each side of the mouth. Further sorting occurs here, and the larger particles are carried to the tips of the palps and then dropped off into the mantle cavity, from which they are removed.

Digestible materials are carried to the deep groove between the two palps. This groove leads directly into the mouth, which lies

between the two 'lips', or ridges, that connect the palps of one
side to the palps of the opposite side. There is no radula, nor
could it be of any use to the animal that feeds only on microscopic particles. The food, entangled in strings of mucus, goes
into the mouth and through a narrow tube, the oesophagus,
to a saclike stomach, which is connected by ducts to a large
DIGESTIVE GLAND. This gland surrounds the stomach and is the

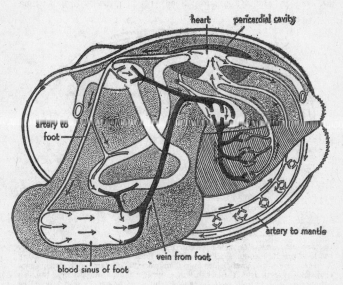

The CIRCULATORY SYSTEM of the clam is an *open* system with blood vessels that supply
and drain irregular channels and sinuses in the tissues. The filling of the large sinus in
the foot helps to produce the swelling of the tip of the foot in locomotion.

main organ of digestion. From the stomach the intestine runs
ventrally, makes several coils through part of the foot, and then
runs dorsally again, passing through the cavity that surrounds
the heart and appearing to pass through the heart itself (actually,
the heart is wrapped around the intestine). The anus opens near
the excurrent opening, and the faeces are carried away in the
outgoing current.

Digestion in fresh water clams is not well understood; most of what
we know about the physiology of digestion in pelecypods is based on

studies of certain marine clams. As we might expect in animals that eat only finely-divided food, the digestion is mostly intracellular. Food from the stomach enters the digestive gland, the cells of which readily ingest and break down solid particles. Protein and fat digestion are exclusively intracellular, and the cells of the gland also absorb carbohydrates. The only extracellular enzyme is the carbohydrate-digesting one, set free in the stomach by the dissolution of the *crystalline style*, a gelatinous rod that lies in a pouch off the intestine and projects into the stomach. The style-pouch is lined with cilia, the beating of which

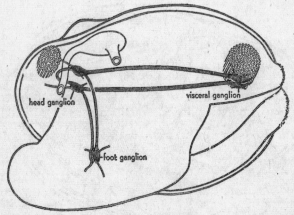

The NERVOUS SYSTEM of the clam has a pair of ganglia for each main region of the body. The head ganglia correspond to the double but fused 'brain' of other animals. They lie on each side of the mouth, joined by a commissure that runs around the oesophagus. They send nerves to palps, anterior shell muscle, and mantle. From each head ganglion a connective runs ventrally to the ganglia which supply the muscles of the foot. Two connectives also run from the head ganglia to the visceral ganglia, which send nerves to the digestive tract, heart, gills, posterior shell muscle, and mantle.

rotate the style and move it forward so that its free end is constantly rubbed against a special portion of the stomach wall. In this way the head of the style is worn away and its material mixed with the stomach contents.

The CIRCULATORY SYSTEM consists of a heart and blood vessels. The heart has three chambers: right and left thin-walled *atria* (singular, atrium), which receive the blood, and a single muscular *ventricle* which pumps the blood. The heart lies in the pericardial cavity, which is lined with an epithelium and filled with fluid.

It represents a remnant of the body cavity or coelom, which is so extensive in annelids (see chap. 19) and other groups. The heart pumps blood both forward and backward through arteries. Anteriorly, it supplies the viscera and foot. Posteriorly, it supplies the mantle. Many of the arteries lead, not into fine capil-

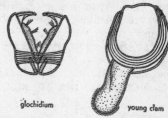

glochidium

young clam

Stages in life-cycle of fresh-water clam. (After Lefevre and Curtis.)

laries, but into irregular channels in the tissues. These channels, called *blood sinuses*, lack the epithelial lining of true blood vessels. From the sinuses the blood flows into veins and then back to the heart. The blood is colourless and contains amoeboid cells.

The EXCRETORY ORGANS lie, one on each side, just below the pericardial cavity. Each is like a tube bent back on itself, with the two parts lying parallel and one above the other. The lower part, or *kidney* proper, has glandular walls. At its anterior end it connects with the fluid-filled pericardial cavity. At its posterior end it is continuous with the thin-walled *bladder* which lies above the kidney and opens anteriorly into a dorsal gill passage. The kidney extracts waste products of metabolism from the blood and from the pericardial fluid. The wall of the bladder is ciliated and maintains an outgoing current.

The body is muscular and performs co-ordinated movements; but the NERVOUS SYSTEM is reduced, as would be expected in so sluggish an animal. There is no head, nor would it be of much use to an animal that lives with its anterior end buried in the mud. Each of the three main regions of the body – 'head', foot, and viscera – has a pair of ganglia; and from the head ganglion two long nerves (connectives) run to each of the other two pairs of ganglia. The SENSE ORGANS are poorly developed. Near the foot ganglia is a pair of hollow vesicles, lined with sensory cells and containing a limestone concretion, which are thought to be balancing organs. A patch of yellow epithelial cells (the osphradium) lies on the visceral ganglia and is thought to be sensitive to chemicals in the water that enters the incurrent opening. The mantle has scattered sensory cells, which are most abundant on the small projections along the edges of the mantle at the openings for the water currents. They probably respond to touch and light. When a clam is irritated, the foot and mantle edges are withdrawn, and the two valves close very tightly – or, as we say, 'shut up like a clam'.

The REPRODUCTIVE SYSTEM consists of a glandular mass, which surrounds the coils of the intestine that lie in the foot, and opens near the external pore of the bladder into a dorsal gill passage. The sexes are separate. The male sheds sperms from the testis

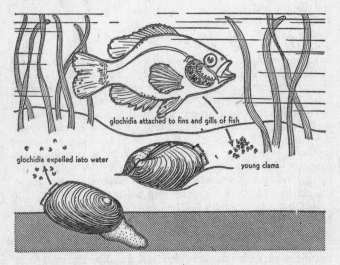

glochidia attached to fins and gills of fish

glochidia expelled into water

young clams

LIFE-CYCLE OF A FRESH-WATER CLAM. The glochidium clamps its valves tightly into the tissues of its host and in some way stimulates them to grow around it, thus forming the so-called 'blackheads' of fish. After 3-12 weeks of parasitic life, the young clam falls off and becomes independent. (Based on Lefevre and Curtis.)

into the outgoing water. They enter the female through the incurrent opening, pass through the pores on the surface of the gills, and reach the interior of the gills, where the eggs are held and fertilization effected. The zygotes develop within the gills to bivalved larvas called GLOCHIDIA (singular, glochidium, 'point of an arrow'). Tremendous numbers of glochidia are produced and expelled into the water, where they slowly sink to the bottom; but most of them die. To develop further, they must, within a few days, become attached to the fins or gills of a fish and live as parasites until they have developed into young clams. Then they drop from the fish and take up the independent life of the adult clam.

In some fresh-water clams there is no parasitic stage; the young develop within the body of the mother. Marine pelecypods shed eggs or sperms into the water, where fertilization takes place. There is first a trochophore and then a veliger larvas (The glochidia of fresh-water clams correspond to the veliger. of marine types.)

CEPHALOPODS

THE most highly organized molluscs are the nautiluses, squids, and octopuses – all marine and members of the class CEPHALO-PODA. The name means 'head-footed', for in these animals the foot, which is divided up into a number of 'arms', is wrapped around the head.

As in gastropods, all degrees of reduction of the shell can be found. While the nautiluses have a large, calcareous, external coiled shell, the squids have only a thin horny vestige of a shell embedded in the mantle, and the octopuses have no shell at all.

The SQUID is one of the most highly developed invertebrates. Some of its structures will be described here to illustrate the ways in which the squid has adapted the molluscan body plan to an active, predatory life.

Unlike most bilateral animals, which are elongated in an anteroposterior direction, the long axis of the squid is dorso-ventral. To compare the body with that of a clam, one would have to place the squid so that the foot was down and the pointed end up. The functional upper surface of a swimming squid is structurally the anterior surface. The functional under surface is structurally posterior. Thus, a squid usually swims with the ventral surface forward, the dorsal surface hindmost, the anterior surface up, and the posterior surface down.

The squid relies for protection not on a heavy shell but chiefly on its ability to leave the scene of danger in a hurry. The SHELL is vestigial and is represented by a feather-shaped horny plate buried under the mantle of the anterior surface. The MANTLE is thick and muscular and has taken on the protective function which in other molluscs is served by the shell. It is also the chief swimming organ. At the dorsal end, its anterior surface is extended into a pair of triangular folds or 'fins', which can be undulated to move the animal slowly and to change its direction of movement. Ventrally the mantle ends in a free edge, the COLLAR, which surrounds the 'neck' between the head and visceral mass. The collar articulates by three interlocking surfaces (ridges which fit into grooves) with the visceral mass and

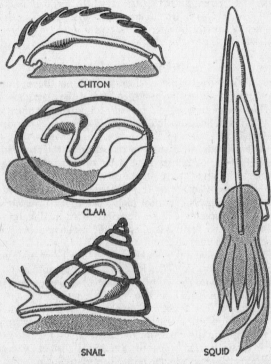

CHITON

CLAM

SNAIL

SQUID

The MOLLUSCAN BODY PLAN has been modified in the various groups. The digestive tract is shaded, the foot is stippled, and the shell is indicated by the heavy black line.

with the FUNNEL, a conical muscular tube that projects beyond the collar on the underside of the head. When the mantle is relaxed, water enters the mantle cavity around the edge of the collar; and when the mantle contracts, the edge is tightly sealed and water is forced out through an opening in the funnel. When the squid is excited, the mantle is contracted strongly, forcibly expelling a jet of water from the funnel. This pushes the animal in the direction opposite to that in which the jet is expelled. When the tip of the funnel is bent backward, the squid darts quickly forward to seize its prey. When the tip of the funnel is directed forward, the animal shoots backward like a torpedo; and this is its usual behaviour in escape. When attacked, it may emit, from a special INK SAC which opens into the funnel near the anus, a cloud of inky material. The 'ink' is thought to serve as a 'smoke screen'; but it has also been suggested that it forms a dark object that distracts the enemy while the squid goes off in another direction.

The 'foot' of the squid is subdivided into the funnel and ten sucker-bearing ARMS which surround the mouth. When the animal is swimming, the arms are pressed together and aid in steering. Two of the arms are different from the rest and can be extended forward to seize the prey with their SUCKERS and to draw it towards the mouth. There it is held firmly by the other arms, while two strong horny JAWS in the mouth kill the prey, biting out large pieces, which are then swallowed so rapidly that the RADULA (which is quite small in the squid) is probably seldom used.

The active life of the squid would not be possible with the slow type of respiration that serves the clam. The contraction and expansion of the mantle provide a steady and effective circulation of water through the mantle cavity, in which lie the two GILLS.

The SHELL of the squid is a thin, horny plate which lies buried under the mantle of the anterior surface.

The CIRCULATORY SYSTEM is also much improved and provides for the rapid distribution of oxygen through the tissues. The blood flows within vessels, which are lined throughout with an epithelium – not into irregular unlined spaces among the tissues, as in the clam. The tissues are permeated with networks of very

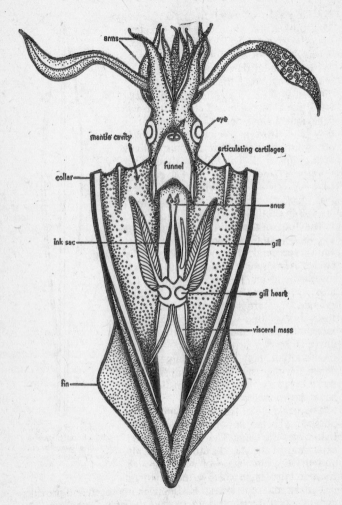

SQUID with mantle slit open along the posterior surface to show the organs in the mantle cavity. Water enters the mantle cavity around the edge of the collar and leaves through the funnel, as shown by the arrows.

small vessels, the CAPIL-
LARIES, through the thin
walls of which gaseous exchanges take place rapidly.
There are separate pumping
mechanisms for blood going
through the gills and that
going out to the tissues. The
deoxygenated blood returning from the tissues enters
two *gill hearts*, each of
which pumps the blood
through one gill. This gives
the blood a fresh impetus,
so that it passes through the
gills at higher pressure.
Freshly oxygenated blood
from the gills enters a single
systemic heart, from which
it is pumped out again to
the tissues.

The NERVOUS SYSTEM of
the squid is very highly developed – in sharp contrast
with that of its slow-moving
relative, the clam. A large
brain encircles the oesopha-

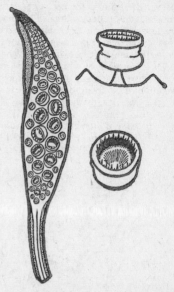

Left, an arm of the squid, showing numerous
suckers. *Right*, *above*, a sucker showing the
muscular stalk by which it is attached. *Right*,
below, sucker showing the toothed, horny
ring with which it is lined.

gus and lies between the eyes. The brain is unusual in that it consists of several pairs of ganglia all fused together, and therefore
has several centres of nervous control, which in lower invertebrates are spread out over the animal. Besides olfactory organs
and a pair of structures that probably aid in balancing, the squid
has two large IMAGE-PERCEIVING EYES. They are remarkably like
the human eye in construction but are developed in quite a different way. When two similar structures having a similar function
appear in two distantly related groups, so that there is no possibility of a common ancestor which could have possessed such a
structure, then the structure must have evolved independently.
Thus the eye of the squid and the eye of man are said to have
arisen by CONVERGENT EVOLUTION.

The two groups of animals which show the best development of eyes, the vertebrates and the cephalopods, also have the most highly developed light-producing organs. In some fishes and squids these organs are amazingly complex, having, besides the photogenic cells which produce the light, lens tissue, reflector cells, and a layer of pigment to screen the animal's tissues from the light.

The occurrence of light-producing organs in two distantly related groups like squids and fish is no more strange than the convergent evolution of the two groups toward the same general type of eye. For BIOLUMINESCENCE, or the production of light by living organisms, is a widespread phenomenon found in bacteria and fungi and in almost every phylum of animals. Its distribution among animals seems quite hit-or-miss, following no special evolutionary lines but accompanying certain ways of life. No luminous fresh-water animals are known, but

Three rows of TEETH FROM THE RADULA of the squid.

animal light is extremely common in marine forms, particularly coelenterates and comb jellies. The 'burning of the sea' at night is caused mostly by luminous protozoa, chiefly flagellates (*Noctiluca, Ceratium, Gonyaulax*). Among luminescent coelenterates we find many jellyfishes (*Liriope*), many hydroids (*Obelia*), siphonophores, scyphozoa (*Aurelia, Cyanea*), and some gorgonians. Practically all the common ctenophores of the North American coasts produce light (*Pleurobrachia, Mnemiopsis* and *Cestus*). Other luminescent invertebrates are ectoproct bryozoans, numerous annelids (polychetes), echinoderms (brittle stars), arthropods (crustacea, fireflies, glow worms), besides many molluscs (mostly cephalopods). Luminous bacteria and fungi emit a continuous light and are responsible for the luminosity of decaying flesh and rotting wood, but most invertebrates shine only when stimulated. In flagellates, medusas, and comb jellies any disturbance of the water, as by a passing boat, will cause the animals to flash; in the case of the firefly (which is really a beetle) there is a definite rhythm of flashing determined by internal stimuli and controlled by the nervous system.

The value of luminescence to living organisms is not clear in most cases. It is thought that light from such forms as luminous bacteria is a by-product of metabolism and has no significance for the life of the organism. It has been suggested that the light organs of squids or fish serve as lanterns, to attract prey or repel enemies; but it is difficult to explain why such annelids as *Chaetopterus*, which passes its whole life in an opaque tube, should be luminous. In the case of the firefly and of certain marine annelids the light does seem to serve as a signal to bring the sexes together for mating.

Only two important steps have been made in our understanding of the physicochemical nature of animal light. Long ago it was shown

THE EYE OF THE SQUID (as well as that of man) is called a 'camera eye' because it is built on the same principle as a camera, which consists of a dark chamber to which light is admitted only through an opening (pupil) in the diaphragm (iris). Behind this opening is a lens which focuses the light on a light-sensitive film (retina).

that luminous wood stops glowing when placed in a container from which the oxygen is removed. Later it was shown that luminescence in the pelecypod, *Pholas dactylus*, is the result of the interaction of two substances which were extracted from the luminous tissues of the animal. If a hot-water extract and also a cold-water extract are prepared and allowed to stand until the light disappears from the cold-water extract, when the two are mixed together light will be produced. The hot-water extract is supposed to contain a substance, *luciferin*, which is not destroyed by heating; the cold-water extract contains an enzyme *luciferase*, which is destroyed by heating. When the extracts are mixed, the luciferin, in the presence of the enzyme, luciferase, becomes oxidized, with the production of light. (Luciferin is at first also present in the cold-water extract; but it soon becomes oxidized, only luciferase remains, and the light disappears. Then, when luciferin is again added, the light reappears until all the luciferin is oxidized.) Similar substances have been demonstrated in fireflies, crustacea, and many other higher invertebrates. But they have not been shown in the lower invertebrates or in bacteria.

Another example of convergence between squids and vertebrates is the development of INTERNAL CARTILAGINOUS SUPPORTS.
The squid has a number of internal cartilages which support
muscles and form interlocking surfaces, but most interesting in
this connection is the large cartilage which encloses and protects
the brain, reminding us of the vertebrate brain case. The squid,
perhaps more than any other invertebrate, has evolved along
the same lines followed by the fast-moving predatory aquatic
vertebrates: large size, streamline shape, rapid locomotion,

LUMINESCENT SQUIDS are mostly deep-sea forms which live in perpet-
ual darkness. It seems reasonable that the *light-producing organs* serve
as lanterns, but this would be difficult to prove. (Based on Chun.)

internal skeletal supports, very efficient respiratory and circulatory systems, large brain, and highly developed sense
organs.

The giant squids are responsible for many of the 'sea-monster' stories.
One of the authentic cases was a squid encountered by the French
battleship, *Alecton*, in the Atlantic in 1860. The monster was 50 feet
long, exclusive of the arms, and 20 feet in circumference at its
largest part. Its weight was estimated at 2 tons. The resistance of such
an animal, even though sick, as this one probably was, can be judged
from this old account: 'The commandant, wishing in the interests of
science to secure the monster, actually engaged it in battle. Numerous

shots were aimed at it, but the balls traversed its flaccid and glutinous mass without causing it any vital injury ... They succeeded at last in getting a harpoon to bite, and in passing a bowling hitch round the posterior part of the animal. But when they attempted to hoist it out of the water the rope penetrated deeply into the flesh, and separated it into two parts, the head with the arms and tentacles dropping into the sea and making off, while the fins and posterior parts were brought on board, they weighed about forty pounds.'

The GIANT SQUIDS are the largest of all invertebrates. (After an old engraving from Figuier.)

BY comparing the clam and the squid, we see that the fundamental body plan of an animal may become so modified in adaptation to a special way of life that many of its structures reflect the kind of life it leads rather than its relationship to its more typical relatives.

CHAPTER 19

Segmented Worms – Nereis

WE frequently refer to the 'average man' or the 'average student'. Biologists who have given some thought to the selection of an 'average animal' have found little difficulty in deciding upon some kind of segmented worm as the animal which would occupy the middle position on a scale of increasing complexity from protozoa to insects or vertebrates. In spite of minor differences of opinion in choosing a particular form, the final vote would go neither to the earthworm nor to the leech, the segmented worms most familiar to everyone, but to some less specialized worm, like the NEREIS.

The Nereids of Greek mythology were sea nymphs, usually represented in the female human form. Their invertebrate name-sakes are marine worms, probably beautiful only to zoologists, but certainly graceful as they swim through the water by gentle undulations of the body. The common nereis of the New England coast grows to a foot or more in length and lives under stones or in temporary burrows in the mud or sand between tidemarks.

The most noticeable feature of the nereis, as of the earthworm, is the ringing of the body, which is not merely external but involves nearly all of the internal structures. The name of the phylum to which the earthworm, the leech, and the nereis belong is ANNELIDA ('ringed'). The ringed condition is more often known as SEGMENTATION, and each ring is called a SEGMENT.

Except for the head and the last segment, all the segments of the nereis are externally alike. They have on each side a pro-jecting appendage or PARAPOD ('side foot'), consisting of

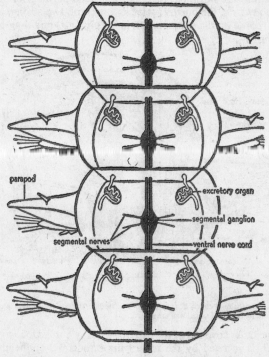

Diagram to emphasize the repetition of parts in the annelid body plan.

flattened fleshy lobes from which protrude bundles of horny BRISTLES. In addition to the undulations of the whole body, a swimming nereis uses its rows of parapods as a series of loco-motory paddles. The bristles are sharp and probably serve a pro-tective function, as well as enabling the animal to obtain a hold on the smooth walls of its burrow.

Although well equipped for swimming, the nereis spends most

of the time in its burrow in the sand, with only the head occasionally protruding above the surface. Gentle undulatory movements of the body create a current through the burrow, bringing the worm chemical stimuli from food organisms located near by and also constantly renewing the water for respiration. In feeding, the worm has been seen to extend the anterior part of the body from the burrow, seize its prey with two strong, horny jaws (borne on the end of an eversible pharynx) and drag it into the burrow.

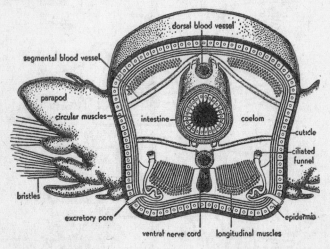

A three-dimensional CROSS-SECTION THROUGH A NEREIS. The coelomic lining is omitted. The muscles of the intestinal wall are shown.

The outer covering of the nereis is a horny but flexible CUTICLE, which is secreted by the underlying ectodermal epithelium, or EPIDERMIS. Beneath the epidermis is a layer of circular muscles, then a layer of longitudinal muscles, and finally a thin lining layer of mesoderm cells (the coelomic lining discussed below). Together these various layers constitute a definite BODY WALL. They run the length of the worm and are divided up by the segmental partitions. More clearly segmental are the bundles of oblique muscles which run in each segment from the mid-ventral line to the parapods, which they move.

To describe one segment of a nereis is to describe nearly the

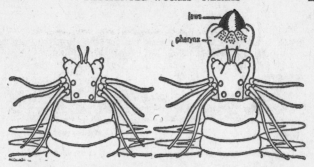

Left, head of the nereis seen from above. *Right*, the pharynx everted.

whole worm. Only the DIGESTIVE SYSTEM shows much differentiation from anterior to posterior ends. The mouth leads into a pharynx, on the inner walls of which are the two large jaws already mentioned. When the pharynx is turned inside out and extended through the mouth, the jaws grasp the food, which is then swallowed by withdrawing the pharynx. Behind the pharynx the digestive tube narrows to an oesophagus, which runs through several segments and into which opens a pair of glandular pouches, the digestive glands. From the oesophagus a long, straight intestine runs the length of the body to the anus in the most posterior segment. The structure of the digestive system presents a definite advance over the condition in the nemerteans in that outside of the digestive epithelium, but in the wall of the digestive tube itself, are thick muscle layers. The contractions of these muscles produce a succession of rhythmic waves of constriction, a type of muscular activity called PERISTALSIS, which push the food along, independently of movements of the whole body.

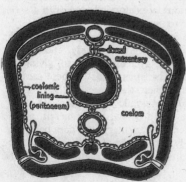

Diagram of an annelid showing the COELOM and its lining.

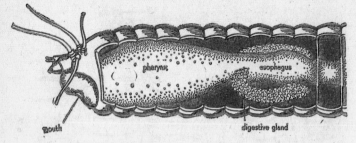

The DIGESTIVE SYSTEM of the nereis is specialized only at the anterior end.

Between the digestive tube and the body wall of the nereis is a definite space, called the *body cavity* or COELOM, which is lined completely by a sheet of mesoderm cells, the *coelomic lining*. One important advantage of the coelom is that it separates the intestine from the body wall and thus permits a freer play of the body-wall muscles. As already mentioned, this also allows the muscles of the digestive tube to push food along independently of the movements of the body. The coelom is filled with a fluid which contains amoeboid cells and many dissolved substances. The coelomic fluid bathes all the internal organs and thus serves a role similar to that of the circulatory system, even though it has no direct connection with that system. The coelom also plays a role in excretion and in reproduction, as we shall see later. Behind the oesophagus the coelom is not a continuous space but is divided up, by the partitions of coelomic lining, into a series of chambers that correspond to the external segmentation.

The coelom arises by the formation of a pair of spaces in the embryonic mesoderm of each segment of the body. These spaces enlarge, and are lined by a thin layer of mesoderm, giving rise to a series of coelomic sacs. The inner walls of these sacs envelop the digestive tube; and where they meet in the mid-line, they form a double layer of coelomic lining, the *mesentery*, which supports the gut above and below. In the nereis, and in many other annelids, the ventral mesentery (the part below the digestive tube) is present only in the embryo. It disappears in the adult, and right and left coelomic spaces are confluent below the digestive tube.

The presence of a coelom is considered of such importance that we often divide animals into two large groups, those with a coelom and

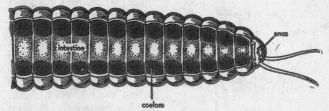

coelom

The intestine of the nereis is not embedded in mesenchyme, as in flatworms and nemerteans, but is surrounded by the coelomic cavity. Most of the middle portion of the worm has been omitted.

those without it, categories which correspond roughly with what we mean when we talk of the 'higher' and 'lower' invertebrates. A space between the digestive tube and body wall occurs in many of the phyla we have studied already. But in such groups as the roundworms and rotifers it has no definite mesodermal lining and therefore cannot be considered a true coelom. In ectoproct bryozoa, brachiopods, arrow worms, phoronidea, and molluscs the body cavity is a coelom. In molluscs it is reduced, for the most part, to the cavity surrounding the heart.

The general type of body structure seen in the nereis – with a muscular body wall separated from a muscular digestive tract by a space lined with mesoderm – occurs in all vertebrates. In man the coelom is divided into an abdominal cavity, a cavity surrounding the heart, and two cavities which contain the lungs. The coelomic lining of animals is also called the 'peritoneum'; and when it becomes infected in man, as from a ruptured appendix, the serious condition that results is known as 'peritonitis'.

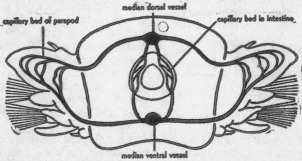

median dorsal vessel

capillary bed of parapod

capillary bed in intestine

median ventral vessel

The main SEGMENTAL BLOOD VESSELS of the nereis are those to the parapods, where the blood is aerated, and those to the intestinal wall, where food enters the blood to be distributed to other tissues.

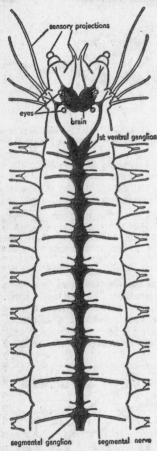

The NERVOUS SYSTEM of the nereis is clearly segmental. All segments are alike, except at the anterior end.

The CIRCULATORY SYSTEM of the nereis does a much better job than the crude apparatus which we saw in nemerteans. The main vessels are a median dorsal vessel, which runs above the digestive system, and a median ventral vessel, which runs just beneath the digestive system. These two longitudinal vessels are connected with each other through the transverse segmental vessels which they give off in each segment. Dorsal and ventral branches of the segmental vessels go to the intestine, parapods, and body wall and there branch and rebranch repeatedly, finally joining each other by way of an intricate network of very fine vessels, the CAPILLARIES. The walls of the capillaries are composed of only a single layer of flattened epithelial cells and are similar in structure to the capillaries of man. Their thin walls permit a rapid exchange of dissolved food substances, nitrogenous wastes, and respiratory gases. Their extensive ramification ensures that substances are delivered almost 'at the door' of every cell and do not have to move long distances by the slow process of diffusion.

Extensive branching does not of itself make a good circulatory system, for, as the name of the system implies, the blood must be in constant circulation. In the nemerteans this is accomplished by an indirect and inefficient method. Waves of muscular contraction passing down the body press on the mesenchyme layer, which in turn pushes against the walls of the blood vessels, forcing the blood along. In the nereis the median dorsal vessel and the lateral branches have muscular walls and are themselves contractile. Rhythmic waves of muscular contraction, of a peristaltic nature like those of the intestine, run forward along the dorsal vessel from behind, driving the blood anteriorly. The blood flows posteriorly in the non-contractile ventral vessel.

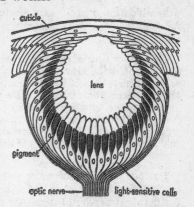

The EYE OF THE NEREIS is advanced over that of a planaria. In addition to the layer of pigmented retinal cells (rods) there is a gelatinous lens which concentrates the light upon the specialized rodlike ends of the sensory cells. The other ends of the cells are continued as nerve fibres which run in the optic nerve to the brain. (Based partly on Kukenthal.)

Besides a good circulatory system to distribute the oxygen after it has entered the blood, large and active animals require an extensive RESPIRATORY SURFACE which is freely exposed to oxygen, either of the air or dissolved in water. In the nereis the amount of body surface exposed to the environment is enormously increased by the thin, flattened parapods, within each of which is an extensive network of capillaries. The capillary beds of the parapods and of the dorsal and ventral body walls lie very close to the surface and, as the blood passes through them, it receives oxygen from the surrounding water and gives up carbon dioxide collected from the tissues. The oxygen-carrying capacity of the blood is increased by the presence of haemoglobin, which is dissolved in the fluid instead of being contained within cells, as in some nemerteans and in vertebrates.

The EXCRETORY SYSTEM is segmentally arranged. A pair of excretory organs lies on the floor of nearly every segment. Each organ consists essentially of a tube which opens at one end by a ciliated funnel into the coelom and at the other end by a pore to the exterior. Wastes extracted from the blood which passes through the excretory organ and also from the coelomic fluid, with which the organ communicates through its internal open funnel, are swept to the exterior by means of cilia lining the excretory tube.

The excretory organs of annelids are called NEPHRIDIA (singular, nephridium). In the nereis the excretory tube is coiled through most of its length, and the coils are compacted within a granular oval mass. The external pore lies at the base of the parapod. The internal end runs forward, passes through the coelomic partition, and opens into the coelomic chamber just anterior to that in which lie the main body of the organ and the external pore. Microscopic particles in the coelomic fluid are wafted into the opening of the funnel by beating of the cilia around its edge.

The concentration of nerve cells into a CENTRAL NERVOUS SYSTEM with a brain and nerve cords marked the advance of the flatworm nervous system over that of coelenterates. In the nereis this tendency toward centralization is carried still farther. The head bears two pairs of eyes and several pairs of projections, which are sensitive to touch and food and other chemicals in the water. With so many sense organs at the head end, there is an increase in the size of the brain. But the brain is only the first and the largest of a series of compact masses of nerve cells, or GANGLIA, which occur in each segment of the body. The head ganglion or brain lies above the pharynx and connects, by a ring of nervous tissue, with another large ganglion lying below the pharynx. From this first ventral ganglion the nerve cord runs posteriorly, and in each segment enlarges into a segmental ganglion, which gives off nerves to the muscles of the body wall and parapods. Each segmental ganglion is like the governor of a state; while the brain, which receives and co-ordinates the various impulses coming from certain of the sense organs on the head, is like the chief executive of a nation.

The primitive brain, as we saw it in the planaria, served chiefly as a sensory relay – a centre for receiving stimuli from the sense organs and then sending impulses down the nerve cord. This is also true of the nereis, for, if the brain is removed, the animal can still move in a co-ordinated way – and, in fact, it moves about more than usual. If it meets some obstacle, it does not withdraw and go off in a new direction but persists in its unsuccessful forward movements. This very un-adaptive kind of behaviour shows that in the normal nereis the brain has an important function which it did not have in flatworms – that of *inhibition* of movement in response to certain stimuli.

The REPRODUCTIVE SYSTEM is very simple. Sex cells are budded off from the coelomic lining in most of the segments. Sexual maturity generally occurs at some definite season of the year, and at this time the worms leave their burrows and swim near the surface. The males are attracted to the females; and as the females burst and shed their eggs in the sea water, the males discharge their sperms. After the sex cells are shed, the worms die.

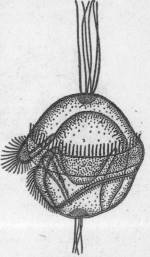

The fertilized egg of the nereis develops into a ciliated larva called a TROCHOPHORE. This larva is of considerable theoretical importance because the same type occurs in several phyla. Few animals seem farther apart in adult structure than a segmented worm

An annelid TROCHOPHORE.
(After Woltereck.)

and a snail. Yet their early stages of development are almost identical, cell for cell; and the trochophores that result are similar in many respects. Beyond the trochophore stage, however, marked differences begin to appear, and the adults are very unlike. The close relationship thought to exist between these two phyla would never have been suspected except for the similarities of their trochophores.

The most remarkable of the embryological resemblances that link the annelids and molluscs is the origin of the mesoderm in the two groups. The early stages of development of certain embryos have been followed so closely that each cell has been numbered and mapped. As a result of this extremely painstaking kind of work it is possible to trace the 'cell lineage' of any portion of the early embryo. The adult mesoderm comes from a single cell (the '4d' cell), which arises in the same way in both annelids and molluscs. This cell divides into a pair, the *primitive mesoderm cells* (shown lying against the wall of the intestine and near the opening of the larval kidney in the diagram of the trochophore). These give rise to two bands of mesoderm, which finally become hollowed out to form the coelom of the adult.

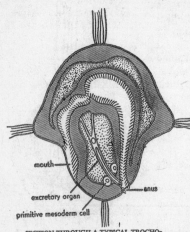

mouth

excretory organ

primitive mesoderm cell

anus

SECTION THROUGH A TYPICAL TROCHO-
PHORE. (Modified after Shearer.)

When the young nereis hatches from the egg membrane, it has already passed the trochophore stage and has three segments with bristles. The 'typical trochophore' usually described is present in only a few annelids, but it is thought to represent the more primitive condition. Its most characteristic structure is a ciliated band about the equator, which serves as the chief organ of locomotion and also directs a food-bearing current towards the mouth, which lies just below. At the upper pole is a group of sensory cells from which arises a tuft of cilia. Some trochophores have a tuft of cilia at the lower pole also. Internally the larva has a complete digestive tract with oesophagus, large bulbous stomach, and a short intestine that opens through an anus at the lower pole. The larval excretory organ contains a flame cell and an excretory tube which opens near the anus.

The development of the annelid trochophore into the adult worm begins with the elongation of the lower region of the trochophore. The elongated region becomes constricted into segments which soon develop bristles. The ciliated bands disappear, and the upper part of the trochophore becomes the head. The young worm then settles to the bottom, takes up a burrowing life, and continues to grow throughout life by the addition of new segments in a region just in front of the last segment.

Various hypotheses have been devised to explain the origin of segmentation, but there is not enough evidence to be able to decide among them. One states that, primitively, all of the organ systems had repeated parts which were spread out over the entire body. The development of crosswise partitions, however they arose, divided the body into segments, each segment receiving representatives of each system. One basis for this hypothesis is the fact that in the planaria and many other animals, structures such as the testes, yolk glands, and the cross-connections between the two nerve cords are repeated along the body, and all that is needed is the development of partitions to produce a segmented condition. Further, in the embryological development of the nereis and other segmented forms, the partitions appear after the basic segmentation is already laid down.

Another hypothesis assumes that each segment represents a subindividual which was produced by asexual budding, as in the planaria, and which failed to detach. In the development of such a flatworm as

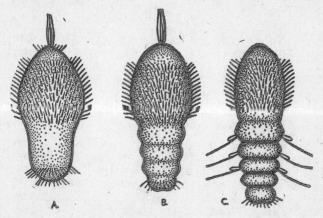

DEVELOPMENT OF AN ANNELID. A, the trochophore is elongating at its lower pole. B, the first segments are indicated, which in C are clearly constricted and already bear the larval bristles. (Modified after Mead.)

Microstomum (see p. 151, Vol. I), chains of as many as sixteen subindividuals form before any break away. A segmented animal, according to this view, is a chain of completely co-ordinated subindividuals. This hypothesis suffers the disadvantage that in a developing annelid the segments do not arise in this way.

Segmentation in one phylum of animals is not necessarily the same as in another. Hence, we cannot expect to devise one explanation for all the events of segmentation, and we cannot point to the common ancestor of all segmented animals, because segmentation probably arose independently in more than one line of evolution.

SEGMENTATION seems to have the same general advantages as the dividing-up of the animal body into cells, namely, there is the possibility for the different segments to specialize in different functions. In the nereis the segments are practically all alike, and this is the primitive condition. In other segmented animals there are varying degrees of specialization, some of which are very extreme.

CHAPTER 20

Earthworms and other Annelids

THE earthworm caught by the early bird is no early worm but one that stayed out too late, for earthworms are nocturnal animals, emerging only at night and retreating underground in the morning. Even at night they usually do not leave their burrows but protrude the anterior part of the body in search of the seeds, leaves, and other parts of plants on which they feed while the posterior end maintains a firm hold on the burrow. These retiring habits have probably contributed to the marked success of an animal that is quite helpless above ground. Since earthworms are adapted to living on so abundant and widely distributed a food as the decaying organic matter of the soil, it is not surprising that they occur in countless numbers in moist soils all over the world. Ever since Darwin made their activities the object of a careful study and concluded that 'it may be doubted if there are any other animals which have played such an important part in the history of the world as these lowly organized creatures', it has been recognized that the work of earthworms is of tremendous agricultural importance.

Earthworms spend most of their time swallowing earth below the surface and depositing it on the surface around the mouths of their burrows in the form of the 'castings' familiar to every-one. In loose soil the burrow may be excavated simply by pushing the earth away on all sides, but in compact ground the soil must actually be swallowed. In moist and rainy weather the worms live near the surface, often doubled up on themselves so that either mouth or anus can be protruded. But in cold weather they plug the opening of the burrow and retreat into its deepest part, which usually ends in a chamber where one or several worms, rolled up together into a ball, pass the winter. In very hot weather, also, they live far from the surface, thus avoiding drying.

The swallowing of earth is not alone a means of digging bur-rows. The soil passed through the digestive tract contains organic materials of various kinds: seeds, decaying plants, the eggs or larvas of animals, and the live or dead bodies of small animals. These are digested, while the main bulk of the soil passes through. When leaves are abundant on the surface the worms drag them into the burrows, and few castings are thrown up. When few leaves are taken in as food, the amount of castings increases.

The *effects of the worms on the soil* are many. The earth of the castings is exposed to the air, and the burrows themselves permit the penetration of air into the soil, improve drainage, and make easier the downward growth of roots. The thorough grinding of the soil in the gizzard of the worm and the sifting out of all stones bigger than those that can be swallowed is the most effective kind of soil 'cultivation'. The leaves pulled into the ground by earthworms are only partially digested, and their remains are thoroughly mixed with the castings, adding organic matter. The excretory wastes and other secretions of the worms also add organic material, enriching the soil for future plant growth. In this way earthworms have helped to produce the fertile humus that covers the land everywhere except in dry and certain other unfavourable regions.

The quantity of earth brought up from below and deposited on the surface has been estimated to be as high as 18 tons per acre per year, or, if spread out uniformly, about 2 inches in 10 years. Seeds are covered and so enabled to germinate, and stones and other objects on the sur-face become buried. In this way ancient buildings have been covered and so preserved, much to the advantage of archaeologists.

Externally the earthworm differs from the nereis in its *adapta-tions to a subterranean life*. As in other burrowing animals, the body is streamlined and has no prominent sense organs on the head or any projecting appendages on the body which would interfere with easy passage through the soil. On each segment are four pairs of BRISTLES, or *setas*, which protrude from four small sacs in the body wall and are extended or retracted by special muscles. The bristles are used to anchor the worm firmly in its burrow, as can be readily discovered by trying to pull one out. But their main function is LOCOMOTION. The worm works its way along by extending the anterior part of the body, taking hold by means of the bristles, and by expansion of the body, then re-tracting the bristles of the posterior region and drawing up the posterior part of the body.

The lack of prominent sense organs on the head does not mean that the earthworm is insensitive to stimuli, but only that there is no concentration of sensory cells into highly specialized organs at the anterior end. As in the nereis, cells sensitive to light, touch, and chemicals occur among the epithelial cells of the epidermis.

The absence of definite eyes, in an animal belonging to a phylum in which well-developed eyes are common, is not unusual. Almost all animals that live in complete darkness have degenerate eyes or no eyes at all; examples are burrowing forms like the earthworm or mole, cave animals like certain fish and crayfish, and nocturnal forms like many beetles.

The *light-sensitive cells* of the earthworm are absent from the ventral surface and are most abundant at the anterior and posterior ends, the regions most frequently exposed to the light. Thought to be organs of *touch*, probably because they occur all over the body, are groups of from thirty-five to forty-five cells, each with a hairlike process which projects through the cuticle covering the surface. Perhaps they are also sensitive to *chemicals* and changes in *temperature*, stimuli to which earthworms respond. *Taste* cells probably occur in the mouth and pharynx, since the worms seem to show definite food 'preferences' – neglecting cabbage if celery is also offered, and passing up celery if carrot leaves are available. The sense of *smell* is very feeble; and the worms are unresponsive to *sound*, which requires a complicated re-ceiving apparatus not found in lower animals. More important for a subterranean animal is the ability to detect *vibrations* transmitted through solid objects. To these, earthworms are extremely responsive. It is said that one way to collect earthworms is to drive a stake into the ground and then move it backwards and forwards, setting up vibrations in the ground, which cause the worms to emerge from their burrows.

The CENTRAL NERVOUS SYSTEM is essentially the same as that of the nereis. A brain (suprapharyngeal ganglion) lying above the pharynx connects by two nerves with a large ganglion (sub-pharyngeal ganglion) lying below the pharynx. These two ganglia send nerves to the sensitive anterior segments and are considered to be the 'higher centres'. The brain is supposed to direct the movements in response to sensations of light and touch which it receives; but if it is removed, the behaviour of the worm is affected little. After removal of the lower ganglion the worms no longer eat, and they cannot burrow in normal fashion. From this first ventral ganglion the double nerve cord runs to the posterior end of the body, enlarging in each segment to a double segmental ganglion from which nerves go to all parts of the segment. Each ganglion serves as a centre which receives impulses coming from

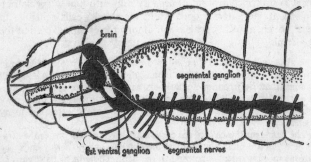

Anterior end of the earthworm, showing the NERVOUS SYSTEM. The digestive system drawn as if transparent.

sensory cells in the skin and sends impulses that result in contraction of the muscles.

The ganglia co-ordinate the impulses so that the longitudinal muscles relax while the circular muscles contract, or the opposite. Without this arrangement the two sets of muscles might only counteract each other's activities, and no movement would result. The smooth muscular waves which pass down the body in the ordinary creeping movements of the earthworm are not controlled by the large ganglia at the anterior end, for almost any sizable piece of an earthworm will creep along as well as a whole worm. The co-ordination is thought to be achieved through impulses relayed from one segment to another by nerve cells in the cord which run from one ganglion to the next. Since there is a certain amount of delay involved in the transfer from ganglion to ganglion in this chain-like succession of connecting fibres, these impulses travel very slowly. If measured as the speed of travel of the waves of thinning or thickening in the body of a moving worm, the rate is only about 1 inch per second. Besides the ordinary creeping movements earthworms can suddenly contract the whole body in response to strong stimulation of any region. If the anterior end is extended from the burrow and receives some unfavourable kind of stimulus, the longitudinal muscles contract as a whole, and the worm disappears into its burrow almost instantly. Such a response requires very rapid nervous transmission, and we do find certain 'giant fibres' in the ventral nerve cord which pass over long distances or even throughout the length of the cord. The speed of transmission in these giant fibres has been estimated at 1.5 yards per second. The speed is about the same in the giant fibres of the nereis, but may be almost 10 yards per second in some of its relatives. These figures seem very low when compared with the rate of nervous conduction in the motor nerves of man, in which impulses travel at about 100 yards per second.

The DIGESTIVE SYSTEM is differentiated into a number of regions, each with a special function. Food enters the mouth, is swallowed by the action of the muscular pharynx, and then passes through the narrow oesophagus, which has on each side three swellings, the CALCIFEROUS GLANDS. These glands excrete calcium carbonate into the oesophagus and in this

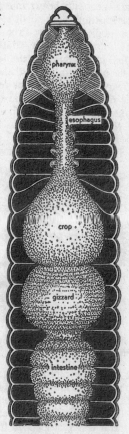

DIGESTIVE SYSTEM of the earthworm. The three pairs of swellings on the oesophagus are the *calciferous glands*.

way dispose of the excess calcium obtained from the various salts present in the food. The oesophagus leads into a large, thin-walled sac, the CROP, which apparently serves only for storage, since the food undergoes no important change and does not remain there very long. Behind the crop is another sac, the

GIZZARD, with heavy muscular walls which (aided by mineral particles and very small stones swallowed by the worm) grind the food thoroughly. From the gizzard the food passes through the intestine, which continues practically uniformly to the anus. In the intestine the food is digested by juices from gland cells of the lining epithelium. The roof of the intestine dips downward as a ridge or fold (the typhlosole), which increases the digestive surface that comes in contact with the intestinal contents. The digested food is absorbed into the blood vessels of the intestinal wall, and from there distributed to the rest of the body.

The CIRCULATORY SYSTEM is very similar to that of the nereis.

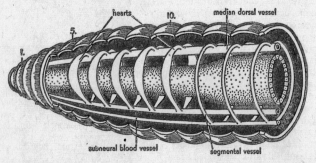

Anterior end of an earthworm, showing the principal blood vessels. Five pairs of HEARTS surround the oesophagus.

A median contractile dorsal vessel, which lies on the digestive tube and accompanies it from one end of the body to the other, is the main collecting vessel. In it the blood flows forward, propelled by rhythmic peristaltic waves. A median non-contractile ventral vessel, suspended from the digestive tube by the ventral mesentery, is the main distributing vessel. In it the blood flows backward and out into branches which supply the various organs. In almost every segment blood flows from the ventral to the dorsal vessel through capillary beds of the body wall, digestive tract, and nephridia. In the region of the oesophagus the dorsal and ventral vessels are connected directly through five pairs of enlarged muscular transverse vessels, the HEARTS, which pump the blood through the ventral vessel. Valves in the dorsal vessel and hearts prevent the blood from backing during irregular contractions.

Branches of the transverse segmental vessels supply blood to the capillary beds of the nephridia and body wall. This blood is then returned to the dorsal vessel. In segments 7–11 this is reversed; the blood flows downward directly from the dorsal to the ventral vessel through the hearts. In front of the hearts the blood in the ventral vessel flows forward to the head; behind the hearts it flows backward and out into the transverse branches. The ventral vessel also sends segmentally repeated branches to the wall of the digestive tube, where the blood becomes loaded with absorbed food. From the intestinal wall the blood returns through paired segmental vessels to the dorsal vessel. Besides draining the body wall and nephridia, the transverse vessels carry blood directly from the subneural vessel to the dorsal vessel. The subneural vessel runs below the nerve cord and supplies it with blood. Two lateral neural vessels (not shown in the diagram) run one on each side of the nerve cord and send branches to the segmental nerves.

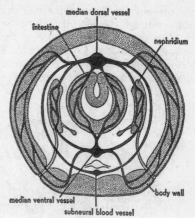

Diagrammatic cross section of an earthworm to show the MAIN SEGMENTAL BLOOD VESSELS.

Earthworms are terrestrial animals, but they have not really solved the problems of land life; they have merely evaded them by restricting their activities to a burrowing life in damp soil, by emerging only at night, when the evaporating power of the air is low, and by retreating deep underground during hot, dry weather. Animals well adapted for land life have a heavy impermeable skin which prevents excessive drying, but it also prevents respiratory exchange through the skin. In such animals oxygen reaches the internal tissues by means of special respiratory devices, such as lungs. Earthworms, on the other hand, breathe in the same way as their aquatic ancestors. That is why they can live for months completely submerged in water, yet will die if dried for a time. The outermost layers of the earthworm are thin and must be kept moist so that RESPIRATORY EXCHANGE can occur by diffusion through the general body surface, which is underlain by capillary networks. Moistening of the surface is accomplished by mucous glands which occur in the epidermis and also by the coelomic fluid which issues from *dorsal pores* located in the mid-dorsal line in the grooves between segments.

The EXCRETORY SYSTEM is like that of the nereis, with a pair of

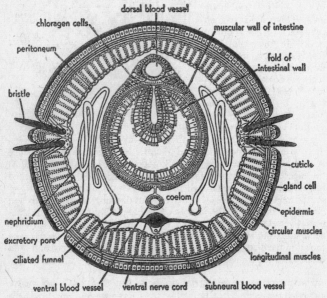

CROSS-SECTION OF THE EARTHWORM. Only two of the four pairs of bristles are shown.

excretory organs, or *nephridia*, in every segment (except the first three and the last). Each nephridium really occupies two segments, because it opens externally by a pore on the ventral surface and internally by a ciliated funnel which lies in the coelom of the segment anterior to the one containing the body of the nephridium and its external pore. The passage of fluid is caused not so much by the cilia lining the nephridial tube as by waves of contraction of the muscles in the wall of that portion of the nephridium which leads to the external pore.

The nephridia are not the only means of excretion in the earthworm. The coelomic lining surrounding the intestine and the main blood vessels is modified into special CHLORAGEN CELLS. Wastes extracted from the blood accumulate in the chloragen cells, which finally become detached and float in the coelomic fluid. Some of the chloragen detritus is removed by the nephridia. Some of it is engulfed by the amoeboid cells of the fluid, which finally wander into the tissues and disintegrate, leaving their wastes as a deposit of pigment in the body wall.

The pigment which thus accumulates in the body wall probably serves to shield the underlying tissues from the light, particularly the ultraviolet, which is very harmful to earthworms. One hour's exposure to strong sunlight causes complete paralysis in some worms, and several hours' exposure is fatal. This is thought to explain the death of many of the earthworms seen lying in shallow puddles after rain. They have not been drowned by the water, as many people suppose, for they can live completely submerged in water. However, during rain the water that fills their burrows has filtered down through the soil and therefore contains very little oxygen. This forces some of the worms to come to the surface, where they are injured by the light and after a time can scarcely crawl. They probably remain in the rain puddles because of the protection afforded by the layer of water. Many of the dead worms seen after rain were no doubt ill beforehand, perhaps as the result of heavy infestation with parasites; their death has only been hastened by the rain.

The organ systems of the earthworm described up to this point have shown little or no increase in division of labour among segments over the condition in the nereis. The nephridia are identical in all segments in which they occur, and the central nervous system is practically the same as that of the nereis. The circulatory and digestive systems of the earthworm show some increase in specialization. Certain of the transverse vessels are enlarged and modified as hearts; and there are (in addition to the pharynx, oesophagus, and intestine, also present in the nereis) two separate regions: the crop for storage, and the gizzard for mechanical breakdown of food. However, the greatest specialization among earthworm segments is found in the reproductive system.

The complexity of the REPRODUCTIVE SYSTEM is an adaptation to land life, where the naked sex cells cannot simply be discharged to the exterior, as in the aquatic annelids, but must in

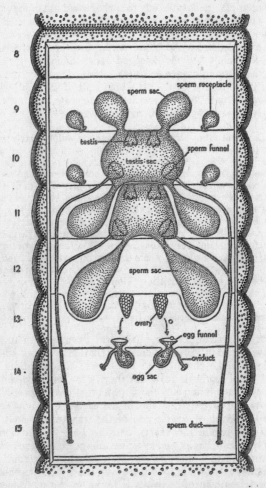

The REPRODUCTIVE SYSTEM of the earthworm is highly specialized. The testis sacs are drawn as if they were transparent, so that the testes and sperm funnels, located within them, can be seen. The sperm ducts actually run beneath the sperm sacs, but have been drawn out at the sides to show their connections clearly. The large sperm sacs in segment 12 press down on the coelomic partition, so that they appear to lie also in segment 13.

some way be protected from drying and other adverse conditions during the development of the young. Earthworms, unlike the nereis, are *hermaphroditic*, each individual having a complete male and female sexual apparatus. This is thought to be an adaptation to sedentary life, which provides relatively few contacts between individuals. Hermaphroditism makes possible two exchanges of sperms, instead of only one, for each meeting of two individuals.

The SEX ORGANS are located in the anterior end of the worm, each organ in a particular segment. The male sex cells are formed

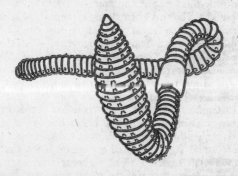

An 'ant's-eye' view of an earthworm with the anterior end lifted to show the four ROWS OF BRISTLES. The thickening of the body wall in the region of the CLITELLUM obscures the external segmentation and also the bristles of that region. The position of the clitellum is definite for each species of earthworm. In *Lumbricus terrestris* it extends over segments 31 or 32 to 37.

in two pairs of testes, located in segments 10 and 11, and each pair is enclosed within a testis sac that communicates with the sperm sac in which the sex cells undergo further development. The mature sperms pass back into the testis sac, into the sperm funnels, and through the sperm ducts to the two male genital openings on the ventral surface of segment 15. Two pairs of small sacs, the sperm receptacles, in segments 9 and 10, open through pores to the ventral surface. During mating they receive the sperms from the other partner. The eggs are formed in a pair of ovaries in segment 13. As they attain maturity, they are shed from the free end of the ovary into the egg funnels situated on the posterior face of segment 13. These funnels lead into the oviducts, which open by two minute pores on the ventral surface of segment 14. The beginning of the oviduct has a lateral pouch, the egg sac, in which ripe eggs are stored. Behind the sex organs is a swollen ring, the CLITELLUM, formed by the thickening of the surface epithelium, which contains great numbers of gland cells. These secrete mucus which, as we shall see, plays an important role in the protection of the developing embryos.

The MATING PROCESS is no simple shedding of the gametes, as in the aquatic annelids. At the sexual season, when the ground is wet following rain, the worms may emerge and travel some distance over the surface before they mate; more often they merely protrude the anterior end and mate with a worm in an adjoining burrow. The two worms appose the ventral surfaces of their anterior ends, the heads pointing in opposite directions. The clitellum of one worm is opposite segments 9–11 of the other, and this is the region of most intimate attachment. Mucus is secreted until each worm becomes enclosed in a tubular mucous 'slime tube', which extends from segment 9 to the posterior edge of the clitellum. When the sperms issue from the male genital openings in segment 15, they are carried backward (in longitudinal grooves which are converted into tubes by the presence of the mucous sheath) to the openings of the sperm receptacles on segments 9 and 10 of the mating partner. Then the worms separate; the egg-laying and fertilization occur later.

The EGG-LAYING starts when the gland cells of the clitellum secrete a mucous ring which glides forward over the body of the worm. As it passes the openings of the oviducts (segment 14), it receives several ripe eggs; and then, as it passes the more anterior openings of the sperm receptacles (segments 9 and 10), it receives sperms which were deposited there previously by another worm in the mating process. Fertilization of the eggs takes place within the mucous ring, which finally slips past the anterior tip of the worm and becomes closed at both ends to form a sealed capsule (sometimes called a 'cocoon'). Within the capsule, which lies in the soil, the zygotes develop directly into young worms and then escape. As in other land animals (and in fresh-water forms as well), there is no free-swimming larval stage comparable with that of marine annelids.

POLYCHETES

THE marine bristle worms, such as the nereis, comprise the largest and most generalized class of annelids, the POLYCHAETA,

GIANT EARTHWORMS are found in tropical regions, especially in Australia. *Megascolecides australis*, in the drawing, may be 11 feet long. It lives in burrows with volcano-shaped openings. (Modified after an old cut from Sterne.)

named from the many bristles borne upon the parapods. They are among the most common animals of the seashore, some living under stones and in the mud in tubes or burrows, while others swim freely in the water, especially during the breeding season.

The FREE-SWIMMING POLYCHETES resemble the nereis in having a well-differentiated head with prominent sense organs, an eversible pharynx bearing horny jaws or teeth, and well-developed parapods. Many of these animals, like the nereis, live in burrows but can leave them and build new ones. Some of them have a special sexual phase, which in the case of the nereids is called a *heteronereis*. This modified sexual animal looks so different from the normal burrowing type that it can scarcely be recognized as belonging to the same species. The eyes are enlarged, the sensory projections shrunken, and the body is differentiated into two distinct parts: an anterior region with unmodified parapods, and a posterior region with parapods that have enormous lobes and flattened, oar-shaped bristles. These changes are associated with the increased activity of the free-swimming sexual form, for the heteronereis does resemble those polychetes which are permanently free-swimming.

In some polychetes the sex cells are formed only in the posterior part of the worm. After undergoing changes in shape and colour this part breaks off, rises to the surface and swims about, shedding the eggs or sperms. In the palolo worm of the South Pacific this takes place at a specific time – just at dawn one week after the November full moon. The sexual pieces rise to the surface in countless millions, and the appearance of the water at this time has been compared to vermicelli soup. Later it appears milky from the eggs and sperms that are discharged. The anterior part of the worm, which remains hidden in some crevice in the coral rock when the posterior piece breaks off, regenerates the missing parts. On the corresponding day of the next year, the regenerated posterior end, laden with sex cells, breaks away.

The natives of the Samoan and other islands are familiar with the habits of the palolos. They consider them a great delicacy and look forward to their breeding season. When the day arrives, they scoop them up in buckets and prepare a great feast, gorging themselves just as we do on Christmas day, knowing that there will not be another treat like it until exactly the same day of the next year. Actually, there is a small 'crop' of swarming palolos a week after the October full moon, but it is too small to interest the natives.

Since the sex cells of animals are capable of being fertilized for only a short time after they are released into sea water, the swarming habits of polychetes, which provide for the simultaneous release of eggs and sperms from a great many closely-approximated individuals, are an adaptation for ensuring the fertilization of the greatest possible number of eggs. In addition to the periodicity of the swarming, some polychetes have other devices which bring about the simultaneous extrusion of eggs and sperms. In the nereids the discharge of sperms is set off by a secretion from the swarming female. In the so-called 'fire worms' of Bermuda the meeting of the sexes involves the exchange of light signals. The worms come to the surface to spawn each month a few days after the full moon at about an hour after sunset. The female appears first and circles about, emitting at intervals a greenish phosphorescent glow which is readily visible to observers on the beach. The smaller male then darts rapidly toward the female, emitting flashes of light as it goes. When the two sexes come close together, they burst, shedding the sex cells into the sea water. Then the spent worms, reduced to shreds of tissue, perish. It has been suggested that the phosphorescent flashes of spawning polychetes were the lights seen by Columbus on the night he approached the New World.

The true TUBE-DWELLING POLYCHETES rarely or never leave their tubes, which may be made of mucus hardened to a parchment-like material, of particles of sand or shells stuck together by a mucous secretion, or of lime laid down on a mucous framework. These worms have degenerate parapods; their heads, though reduced, are provided with long brilliantly coloured tentacles which protrude from the opening of the tube.

Some forms also have extensible gill filaments, full of circulating blood, which serve as respiratory organs. The anterior part of the digestive tube is not eversible, and there are no jaws. These worms feed on minute animals or plants which are carried towards the mouth by rows of cilia on the tentacles.

The BURROWING POLYCHETES have reduced heads and parapods as in the tube-dwelling types. But they have an eversible pharynx, which, in the common lugworm *Arenicola*, is covered with minute papillas and, in addition to its use in feeding, serves as the chief organ of locomotion through the sand. Arenicola feeds like an earthworm, passing large quantities of sand through the digestive system to obtain the organic matter mixed with it. The castings can be seen on sandy beaches when the tide is out.

OLIGOCHETES

THE second largest class of annelids is the OLIGOCHAETA ('few bristles'), of which more than four-fifths are the earthworms and the rest are mostly small or minute worms which occur in the soil or in fresh water. (This last type is extensively used as fish food and is familiar to breeders of pet fish. A hydra can be seen eating a fresh-water oligochete in the photographs in chapter 7.)

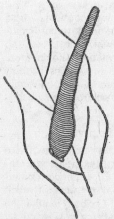

The oligochetes differ from polychetes in several important respects. There are no parapods, and the bristles emerge from pits in the body wall. Whereas the polychetes have separate sexes and the sex cells are budded off from the coelomic lining in numerous segments, the oligochetes are hermaphroditic and the sex cells are produced in special organs which occur only in certain segments. Because both groups have clearly marked external segmentation, share many annelid structures, and bear on each side of every segment two separate bundles of bristles which are moved by muscles and serve in

LAND LEECHES live in tropical forests, attached to leaves by the posterior sucker, with the body held erect and prepared to fasten on to the first mammal that comes within reach. These leeches occur in such numbers that horses are sometimes driven wild by them and men suffer serious loss of blood. (After a photograph by Heinrich.)

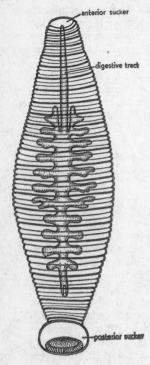

A LEECH. The digestive tract has large side pouches which greatly increase its capacity for blood. The surface of the leech is thrown into folds, of which only certain ones (indicated by heavy lines in the drawing) correspond to segments. This animal lives in the gill chambers of a fish. (After Hemingway.)

locomotion, they are sometimes grouped together as the *Chaetopoda* ('bristle-footed'). These resemblances may be the result of descent from a common ancestral group which was more primitive then either polychetes or oligochetes. But if one of the groups is derived from the other, the polychetes are certainly the more primitive group and the oligochetes the derived group.

ARCHIANNELIDS

FORMERLY both polychetes and oligochetes were thought to have evolved from a supposedly primitive group of annelids, the class ARCHIAN-NELIDA ('primitive annelids'). Now the best opinion considers these worms to be not primitive but *simplified* annelids, which have lost the external segmentation and in most cases also the parapods and bristles. Many of them are ciliated, a juvenile character retained from the trochophore larva.

LEECHES

THE leeches, class HIRUDINEA, have no bristles; and the external segmentation of the body does not correspond with the internal segments, of which there are fewer. The body is solid, the coelomic spaces being crowded out by the growth of connective tissue. There is a definite number of segments, for leeches do not add them on throughout life, as do other annelids. At each end of the body is a sucker, the posterior being much larger than the anterior, which has the mouth in its centre. Despite all the modifications the leeches are closely related to the oligochetes and are probably derived from them. There is no character of leeches which is not present in at least some degree in some oligochete.

Most leeches lead a semiparasitic life, sucking the blood of vertebrates, although some of them have lost this habit and feed on small animals. They show some of the adaptations for parasitism that were noted among flatworms, namely the development of clinging organs, the suckers, and the extreme complication of the reproductive system. On the other hand, since they need to swim about to locate victims, they are less modified than flukes or tapeworms and have eyes. In sucking blood, a leech attaches to some vertebrate by the posterior sucker, applies the

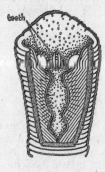

The HEAD OF A LEECH, cut away to show the three sawlike teeth with which it makes a wound. (Modified from Pfurtscheller.)

The three teeth of a leech inflict a Y-shaped wound. (After Reibstein.)

anterior sucker to the skin, makes a wound, often with the aid of little jaws inside the mouth, fills its digestive tract with blood, and then drops off, remaining torpid while digesting the meal. Large blood meals are few and far between, but the digestive tract has lateral pouches which hold enough blood to last for months. The salivary glands of leeches manufacture a substance called *hirudin*, which prevents the coagulation of the blood while the leech is taking its meal. For this reason a wound made by a leech continues to bleed for a long time after the leech has detached itself.

CHAPTER 21

A Missing Link – Peripatus

IF we could find an animal clearly intermediate in structure between two modern phyla, we would have good evidence that the two phyla are closely related. That such an animal has never been found is not surprising. Indeed, it would be more remarkable if the very form which at some remote time in the past gave rise to two stocks, now represented by two modern phyla, had also persisted unchanged through the ages. We have fossil records to show that certain species have remained unchanged for very long periods of time, but none are so old that they trace back to the time before all the modern phyla had evolved. Therefore, we often speak of these missing ancestral forms as 'missing links'.

An animal that comes closer than any other to being the 'missing link' between any two phyla is the PERIPATUS, member of the small phylum ONYCHOPHORA, the name of which means 'claw-bearing' and refers to the curved claws on the feet. The peripatus is a rare animal, found in moist places under logs in the tropical forests of Australia, Africa, Asia, South and Central America, and a few other regions. Its occurrence only in local regions in such widely separated parts of the world suggests that it was probably a more successful and widespread form in the past but is now gradually disappearing.

A peripatus looks much like a cater-pillar, 2 or 3 inches long, with soft velvety skin and many pairs of legs. While many of its structures are like those of the phylum Arthropoda, to which the caterpillars belong, the peripatus also has many similarities to the annelids and, of course, some special features of its own. As already pointed out, neither the peripatus, as we find it now, nor any other living animal could be ancestral to any group as old as a phylum; but there is little doubt that the peripatus is a descendant from a line which branched off close to the primitive annelid-arthropod stock.

LEG of a peripatus showing the arthropod-like claws.

Unlike typical annelids and arthropods, the peripatus shows no external segmentation, though there is a pair of legs for each internal segment of the body. The legs end in claws, which superficially resemble those of arthropods, but differ from arthropod legs in that they are not divided into joints.

The outer covering is a thin CUTICLE like that of annelids, although it is ridged and covered with microscopic projections which give it a *velvety texture* unknown in other animals and which appear to prevent the cuticle from being readily wetted by water. Beneath the epidermis which secretes the cuticle are layers of muscles, as in annelids. The body wall of arthropods is somewhat different, having a heavy outer covering and no continuous layer of muscle.

The peripatus usually comes out at night; and, though it has a pair of simple annelid-like eyes, it feels its way about by means of two sensory projections, or antennas, on the head. When attacked, it gives off a slimy secretion from a pair of glands which open on two projections, the oral papillas. It feeds on small insects and other animals by means of a pair of horny cutting jaws. Each of the three pairs of head appendages (antennas, oral papillas, and jaws) occurs on one of the three segments which compose the head. The fusion of segments, particularly at the head end, is characteristic of the most highly developed segmented animals; and the THREE-SEGMENTED HEAD of the peripatus is thought to indicate a condition midway between that of annelids and arthropods, since the latter have a six-segmented head.

The INTERNAL ANATOMY is a mixture of annelid-like and arthropod-like structures. The digestive tract is simple and not particularly distinctive. The circulatory system is like that of arthropods. A long contractile dorsal vessel, the heart, extends the length of the body and has along its sides a pair of openings for each segment of the body. As in arthropods, there are no definite vessels to return the blood to the heart. After leaving the vessels that carry blood away from the heart, the blood flows into large spaces in the tissues and finally collects in the space surrounding the heart, into which it enters through the paired heart-openings. The coelom is practically obliterated by the growth of connective tissue in which the blood spaces occur, and this is typical of arthropods. But the most arthropod-like character of all is the respiratory system, consisting of AIR TUBES (tracheal tubes) which open from the external surface and extend throughout the body, piping air directly to the tissues. Although such structures occur nowhere else in the animal kingdom except in terrestrial arthropods, they are thought to have arisen independently in the two groups and not to be evidence that the Onychophora are related to arthropods.

The most annelid-like character is the EXCRETORY SYSTEM. This consists of segmentally arranged pairs of coiled tubes which open by external pores at the bases of the legs. They resemble the excretory organs (nephridia) of annelids. The inner end of each organ opens into a very small coelomic sac, from which wastes are collected, presumably entering there by diffusion from the large internal blood space. Cilia in the tube sweep the wastes out through the external pore.

The NERVOUS SYSTEM is primitive. From the brain in the head run two widely separated ventral nerve cords which show small thickenings in each segment. Some annelids have widely separated cords, but even the primitive ones have segmental ganglia that are larger than those in the peripatus.

The REPRODUCTIVE ORGANS are ciliated, as in annelids; cilia do not occur anywhere in arthropods. The eggs are fertilized within the body of the female. In some species eggs are laid, but in most forms the eggs develop within the female, and the young are born fully developed. As internal fertilization and development of the eggs are adjustments to land life and have evolved independently in terrestrial animals of many phyla, they have no special significance for the relationships of the peripatus.

THE existence of an animal with structures peculiar to two different phyla is a situation which follows naturally from what we know of the continuous nature of the process of organic evolution. But it creates difficulties in classification. The problem has been solved temporarily by placing the peripatus group in a phylum by itself. Still, there are some zoologists who think that the animal is definitely an annelid and that its arthropod-like characters have arisen independently in the two groups. Others feel that these curious animals should be made one of the classes of the arthropods, with which group, they maintain, it belongs. That such a controversy exists makes the peripatus the best living candidate for the title of 'missing link'; or, since it is not missing, perhaps we should call it a 'connecting link'. It suggests what the intermediate stage between two phyla might have been like, although the picture is much modified by the fact that the peripatus has undergone considerable evolution since the time it first branched from the primitive annelid-arthropod stock.

CHAPTER 22

Jointed-legged Animals

IN human society 'success' is commonly expressed in terms of the amount of money a man controls or the level of esteem which he occupies in the minds of his associates. But when we talk of the 'biological success' of man or of any other animal as a species, we have in mind very different criteria.

The animal groups which we judge to have attained the greatest 'biological success' are those which have the largest numbers of species and of individuals, occupying the widest stretches of territory and the greatest variety of habitats, consuming the largest amount and kinds of food, and most capable of defending themselves against their enemies. By these standards the phylum which occupies first place among the animals (vertebrate and invertebrate) is the phylum ARTHROPODA.

More species of arthropods have been described than of all other kinds of animals put together. Of the million or so known species of animals, over three-fourths are arthropods. The MAJOR CLASSES of arthropods are the *crustaceans* (crayfish, lobsters, shrimps, crabs, water fleas, barnacles), the *centipedes*, the *millipedes*, the *arachnids* (spiders, scorpions, ticks, mites), and – by far the largest class of all – the *insects*. Only a few insects have been able to invade the ocean, but the group is extremely abundant in fresh water and on land. In temperate regions insects cannot compete with the warm-blooded vertebrates during the winter; but in the tropics, where they suffer no handicap, they are dominant at all seasons.

Some arthropods are beneficial to man, providing food or some valuable service. Others do untold damage, destroying crops, undermining wooden buildings, and transmitting diseases.

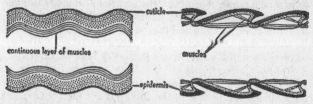

BODY WALL OF ANNELID AND ARTHROPOD CONTRASTED. In annelids the cuticle is thin and the epidermis is underlain by heavy layers of circular and longitudinal muscles. In arthropods the cuticle is heavy; the muscles occur in separate bundles; there is no continuous layer.

Parts of the world's most fertile regions are closed to man by the presence of disease-bearing arthropods. And where they do not exclude him althogether, they are man's chief competitors for food and shelter.

THE ARTHROPOD BODY PLAN may be roughly described as an elaboration and specialization of the segmented body plan of annelids. Primitive arthropods are composed of a series of similar segments bearing similar appendages. But in the higher types almost every segment of the body has a somewhat different structure and function. The outer layer, or CUTICLE, very thin in annelids, in arthropods is usually a heavy layer which serves as a *protective armour*. It is non-living, but is secreted by the underlying epithelium, and is composed of several different substances, each of which contributes some useful property. Horny outer coverings occur in many groups of animals (for

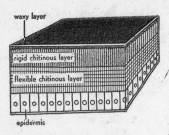

BODY WALL of an arthropod.
(Based on several sources.)

example, the covering of the obelia colony or the cuticle of annelids), but in no case are they used so effectively or produced in so great a variety of structures as in arthropods. Made of the cuticle, in whole or in part, are outer protective coverings, biting jaws, piercing beaks, grinding surfaces, lenses, tactile sense organs, sound-producing organs, walking legs, pincers, swimming paddles, mating organs, wings, and innumerable other structures found among the highly-diversified insects. This horny material is to the arthropods what steel is to civilized man, and it is partly due to the possession of this hard cuticle that the arthropods owe their success.

The surface of the cuticle is a thin waxy layer which makes it *waterproof*. Under this is a heavy layer composed of a protein and of CHITIN, a horny flexible substance which is the most characteristic component of the cuticle, if not the principal one, and which provides *elasticity*. Wherever the cuticle is relatively rigid, as it is over most of the surface of an arthropod, there is a third layer, which lies between the other two. It is formed by the infiltration of the upper part of the chitinous layer with the substance of the waxy layer and other hardening materials (which in crustaceans are mostly lime salts). The middle layer is responsible for the *rigidity* of the cuticle, the property which makes it so effective as a protective armour. Since the hardening occurs in definitely limited areas, between which the cuticle remains as flexible membranes, or joints, the outer covering of arthropods provides *protection* without sacrificing *mobility*. This is what makes it so superior to the armours of such animals as the snails and clams, which have heavy cumbersome shells that limit movement.

Since the rigid cuticle furnishes a supporting framework for the tissues within and provides a surface for the attachment of muscles, it is appropriately called an *exoskeleton*. And though the chitin is responsible for the elasticity, rather than the rigidity, of the cuticle, in order to distinguish it from the external supports of other animals we usually call it a CHITINOUS EXOSKELETON.

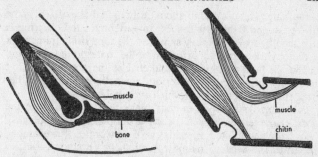

Diagram contrasting skeletal and muscular systems of vertebrate and arthropod. *Left*, part of a vertebrate limb, showing that the bones lie internally and have muscles attached to their outer surfaces. *Right*, part of an arthropod limb, showing that the cuticle lies externally and has muscles attached to its inner surface.

In sharp contrast to this kind of framework is the endoskeleton of vertebrates, which lies on the inside and is surrounded by the soft fleshy parts. We can imagine how it might feel to be an arthropod by mentally putting on an iron suit of armour which adheres closely to the skin, and then thinking of our bones being eliminated and our muscles being attached instead to the iron armour.

To their chitinous exoskeleton arthropods owe their ability to live on land. LAND LIFE requires, among certain other adjustments, a relatively *impermeable outer covering* to prevent drying

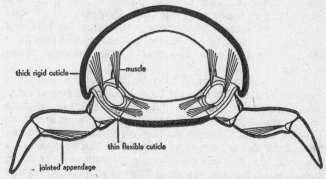

Three-dimensional cross-section through the EXOSKELETON of an arthropod to show that it consists of hardened plates, joined by more flexible membranes of cuticle, and serves as a place of attachment for muscles. (After Snodgrass.)

of the watery tissues within and a fairly *rigid framework* of some kind to support the soft tissues. In vertebrates the covering is furnished by scales or heavy skin, and the supporting framework is a bony skeleton. In arthropods the waterproof and rigid cuticle fills both requirements and enables the group to exploit the land with practically no serious competition from the other invertebrate groups, most of which are largely aquatic.

The name Arthropoda means 'jointed legs' and refers to the most characteristic structures of arthropods. To distinguish them from the jointed appendages of vertebrates (for example, the arms and legs of man) we call them CHITINOUS JOINTED APPENDAGES. The various appendages of arthropods have a specialized structure which adapts them to some particular function. This increases efficiency but sacrifices the versatility possessed by the more generalized hands of man.

LIKE the nereis, which has a pair of swimming flaps (parapods) on nearly every segment of the body, the arthropods also have, typically, a pair of appendages to every segment. In the embryos of both groups the appendages arise in a similar way from similar structures, and hence are said to be 'homologous'. The principle of HOMOLOGY is the basis of our scheme for determining animal relationships. Thus, if two animals have similar structures which develop in the same way from corresponding embryonic parts, the animals are judged to be closely related. The more similar the structures and their mode of origin, and the greater the numbers of such structures, the closer their relation. In other words, we assume that the homologous structures of two different animals have come, by a process of gradual modification, from the same or a corresponding part of some remote common ancestor.

Not all structures which resemble one another indicate a common evolutionary origin of the animals which possess them. Many are only superficially alike, being adapted to the same environmental conditions; and such structures arise in entirely different ways in the embryos. They are similar in function but not in basic plan or mode of origin, and are said to be ANALOGOUS. The wing of a bird and that of a bee are both used for flying, though one is made of feathers and the other of chitin, and they do not develop in the same way. They are analogous, but not homologous. On the other hand, the wing of a bee is homologous to that of a dragonfly or a cockroach. In all three insects the wing is essentially the same and arises from a corresponding part of the embryo. In this case the homologous organs, all used for flying, are also analogous. Sometimes homologous structures have different functions: the legs of a bee are used in walking, while those of a water beetle are adapted for swimming.

When corresponding structures in different segments of the *same* animal are considered, we say that they are SERIALLY

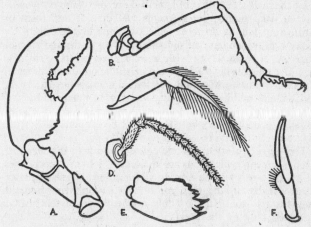

The JOINTED APPENDAGES of arthropods, originally chiefly for locomotion, have been modified for a great variety of functions, even in the same animal. They are all homologous but not analogous. A, pinching leg of a lobster. B, walking leg of a grasshopper. C, swimming leg of a water beetle. D, sensory antenna of a honeybee. E, chewing jaw of a cockroach. F, mating organ of a male lobster. (After various sources.)

HOMOLOGOUS. One parapod of a nereis is serially homologous to any other. The front leg of a bee is serially homologous to the third leg, and both legs are serially homologous to the antennas and the jaws, which are modified segmental appendages and arise from corresponding parts of their respective segments. On the other hand, the eyes or wings arise in a different way and, therefore, are not homologous to the jaws, antennas, and legs.

IN the insects and in many other arthropods the segments are grouped into THREE BODY REGIONS: head, thorax, and abdomen.

While the head always consists of six segments, the thorax and abdomen are composed of different numbers of segments and are not comparable in the various groups of arthropods. The thorax of a crustacean does not correspond to the thorax of an insect; they are similar only in that they both represent the middle region of the body. The head and thorax may be fused, as in the lobster, or the abdomen may be much reduced as in crabs; but in most arthropods the total number of segments is much smaller than that of annelids. Further, adult arthropods have a fixed number of segments and do not add them on throughout life, as do most annelids. The same is true of the vertebrates, which have a fixed number of segments and a fusion of many of them. If we wish to generalize, we may say that primitive animals have a large and indefinite number of repeated but similar parts, while more specialized animals have a smaller and definite number of repeated parts with much division of labour among them, or they have the repeated parts fused into compact masses or organs. (For a specific example of this, see diagrams of the nervous systems on p. 272).

The HEAD of every arthropod consists of exactly *six segments*. Each segment typically bears a pair of jointed appendages, which are sensory or have to do with feeding. The segments are clearly visible in the embryo; but as development proceeds they fuse, so that, in the adult, segmentation of the head is indicated only by the presence of the several pairs of appendages. The head also bears a pair of compound eyes in the primitive arthropods, crustaceans and insects; the others have only simple eyes or clusters of simple eyes. Most insects have simple eyes in addition to the compound eyes.

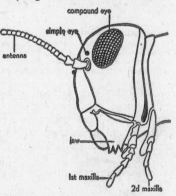

Knowledge of the correspondence between appendages and segments comes from the study of embryology. The following description applies in general to the crustaceans,

HEAD OF AN INSECT, showing structures characteristic of the heads of most arthropods. (Modified after Snodgrass.)

centipedes, millipedes, and insects but not to arachnids (see chap. 23). The first segment never has an appendage. The second bears a pair of feelers, or *antennas*. The third has a *second pair of antennas* in crustaceans, but lacks a segmental appendage in insects. The fourth has the *jaws*, which usually serve for biting but may be modified for other methods of feeding. The fifth and sixth segments each bear a pair of *maxillas*, accessory jaws which aid in feeding, particularly in handling the food and in holding it to the mouth. The mandibles and maxillas are referred to collectively as 'mouth parts', and are frequently very highly modified

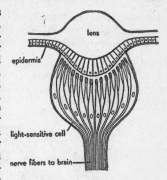

SIMPLE EYE of a spider. All the sensory cells have a single lens, made of cuticle and secreted by the underlying epidermis. (After Hentschel.)

The EYES of arthropods are composed of visual units, each of which is a bundle of cells consisting of two functional parts. The first is represented by *refractive bodies* which transmit the light rays and condense them upon the light-sensitive cells. The cuticle which covers the surface of the body is transparent and usually much thickened to form a lens over the surface of the eye; and there are one or more additional refractive structures within the eye. The second part, or *retina*, lies deeper and is composed of a layer of light-sensitive cells continued at their lower ends into nerve fibres which enter the central nervous system. A SIMPLE EYE has a single light-condensing apparatus for all the sensory cells. A COMPOUND EYE is composed of hundreds or thousands of units, each with its own light-condensing apparatus. This kind of eye is unique to arthropods and (except for the 'camera' eyes of certain molluscs described on p. 225) is the most highly developed of invertebrate eyes. It does not give as sharp an image as the 'camera' eye of man; perhaps the arthropod sees something a little worse than a newspaper photograph as it would look to us under a magnifying glass. However, some insects must have fairly good images, for they have been seen attempting to extract nectar from flowers on wall paper. In any case, arthropods react not so much to details in an image (as we do) as to *motion*. Since the movements of objects are recorded successively in every unit, the compound eye is admirably adapted for detecting the slightest movement of prey or enemy.

Each unit of the compound eye is isolated optically from its neighbours by a screen of pigment cells, so that only a narrow band of parallel rays enters each unit; and it is thought that the image thrown on the retina is a MOSAIC composed of as many points of light as there are units. Most compound eyes are adapted to see in dim light by a migra-

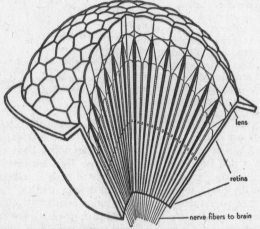

The COMPOUND EYE of an insect is composed of many visual units, each with a small bundle of light-sensitive cells and a lens. The lens is made of cuticle and is secreted by specialized epidermal cells. The light-sensitive cells are also thought to be modified epidermal cells. (Partly after Hesse.)

tion of the pigment, leaving the sides of the visual units exposed. In this case each unit throws on the retina an image of a larger part of the visual field, and the adjacent images overlap somewhat. Such overlapping images are not as sharp, but enable the animal to see in dimmer light, since they do not waste the light which enters obliquely and is therefore absorbed by the pigment in the mosaic type of vision.

The THORAX of arthropods has different numbers of segments in the various groups. In the crustacea, which are more primitive, there are often numerous segments, with appendages for feeding and walking. In the insects, which are more specialized, the thorax is composed of three segments, each of which bears ventrally a pair of LEGS and dorsally, on the second and third

segments each, a pair of WINGS. The
thoracic appendages of arthropods are
most often used for walking, but may
also serve other functions. In the lobster
one pair is modified for grasping; and in
the honey-bee the legs, though used for
walking, are highly modified for collecting pollen (see p. 318).

The ABDOMEN may or may not have
appendages. In the crustaceans, such as
the lobster, there is a pair of appendages
on every abdominal segment; but in
higher forms these have been lost, until
in the insects there are practically no
abdominal appendages homologous
with the appendages of the other segments, except the egg depositing structures on the most posterior segments.

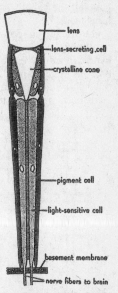

In the nereis we saw a clearly segmental arrangement of parts, both
external and internal – with only the
beginnings of fusion and specialization
at the head end. Every organ system had
a representative in each segment which
provided for the local needs of the
segment. In the arthropods we have
already seen modifications of the primitive external segmentation in the fusion

SINGLE UNIT OF A COMPOUND-
EYE. The lens and crystalline
cone are refractive bodies.
The light-sensitive cells are
surrounded by a screen of
pigment cells, which exclude
oblique rays of light from
adjacent units.

of the head segments and the specialization of the various appendages. The INTERNAL SEGMENTATION is even more modified. Some
of the organ systems consist of single large organs which serve
the whole body, and segmentation is clearly apparent only
in the repeated branches of the circulatory and respiratory
systems and in the ganglia and segmental branches of the
nervous system.

The NERVOUS SYSTEM consists of a dorsal brain which connects,
by a ring of nervous tissue encircling the digestive tract, with
the first ganglion of the ventral nerve cord. In primitive arthropods this system can hardly be distinguished from that of
annelids. In higher arthropods there are all stages of condensa-
tion of the ganglia, reaching a peak in certain animals which

have all the ganglia of the thorax and abdomen fused into one large mass.

One might suppose that an animal that lives encased in a non-living cuticle as heavy as that of arthropods would be handicapped in establishing connections between the central nervous system and the external environment. On the contrary, the cellular and cuticular layers of the body wall of arthropods have been modified to form highly specialized SENSE ORGANS

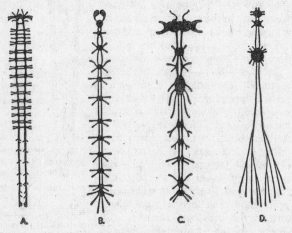

NERVOUS SYSTEMS of various arthropods, showing fusion of the ganglia. A, primitive crustacean. B, caterpillar. C, honeybee. D, water bug. (After several sources.)

of a variety greater than that found in any other phylum. The *eyes*, sensitive to light, and the *antennas*, sensitive to touch and to chemical stimuli, have already been mentioned. Some arthropods also have *balancing organs*, composed of sensory pits containing hard particles, and *auditory organs*, which have a flexible membrane stretched across an opening in the hard cuticle. In addition, the surface of the body is covered with a variety of *sensory bristles*, 'hairs', spines, scales, and pits. The simplest of these is a bristle formed by a hollow outgrowth of the cuticle and connected with a sensory cell which extends to its base. The bristle articulates with the cuticle, and any mechanical stimulus which moves the bristle sets up an impulse in the sensory cell with which it connects. Certain small and slender bristles which are not movable and have thin and permeable walls are assumed to be among the *receptors of chemical stimuli* (taste and smell).

The COELOM of arthropods is practically obliterated. It appears in the embryo as a series of cavities in the mesoderm, but in the adult is represented chiefly by the cavities of the sex organs.

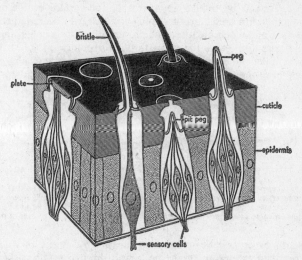

Portion of body wall of an insect to show several kinds of SENSE ORGANS. The *bristle* responds to touch. The *peg* is made of thin cuticle and is therefore thought to be a receptor for taste or smell; the same is true of the *pit peg*, which lies in a cavity that opens to the surface through a pore. The *plate* (without a pore) may be a receptor for chemical stimuli, but its exact function is not known. (Combined from McIndoo and Snodgrass.)

The apparent body cavity of the adult is not a coelom at all, but a large 'blood cavity' which forms part of the circulatory system.

The CIRCULATORY SYSTEM has evolved in the direction of simplification. The *heart* is a pulsating tube which lies dorsally. The arteries lead not into capillaries but into the 'blood cavities' in the mesenchyme throughout the body. In these spaces the blood bathes the various organs. Then it returns to a large cavity surrounding the heart and enters the heart through paired

openings in its sides. Since the blood is not at all times confined
within blood vessels, this type of system is called an OPEN
CIRCULATORY SYSTEM, in contrast to the *closed* systems of such
groups as annelids and vertebrates.

The great adaptability of the arthropod body wall is further
emphasized by the structures concerned with RESPIRATION.
Most terrestrial arthropods have a system of branching *air
tubes*, formed by tubular ingrowths of the surface ectoderm.
The ectoderm secretes an inner lining of cuticle, which strengthens
the walls of the delicate tubes and prevents them from collapsing.
Air enters and leaves the tubes through openings on the sides
of the body and is piped directly to the tissues, partly or almost
completely replacing the respiratory function of the circulatory

Left, CLOSED CIRCULATORY SYSTEM of annelids; the blood is confined within blood
vessels. *Right*, OPEN CIRCULATORY SYSTEM of an insect. The only blood vessel is a pulsating tube, the heart. Blood flows forward through the heart and then passes out of the
open anterior end into the tissue spaces and large blood cavities, eventually returning
to the heart through paired openings in its sides. In the less specialized arthropods the
vessels are more extensive, arteries leading from the heart to the main regions of the
body.

system. Most aquatic arthropods breathe by means of *gills*,
thin-walled extensions of the body wall through which carbon
dioxide and oxygen pass readily.

The cuticle also forms an important part of the DIGESTIVE
SYSTEM. The ectoderm turns in at the mouth and anus and lines
the anterior and posterior regions of the digestive tube with the
cuticle which it secretes. In the anterior region the cuticle may
be produced into hard teeth for grinding up the food. In the
insects and many other arthropods the anus serves also as
the exit for nitrogenous wastes, since the EXCRETORY ORGANS
are tubules which open into the digestive tube.

Aquatic arthropods range in size from minute crustaceans
like the 'water fleas' to monster crabs measuring up to 12 feet
across their long, spindly, outspread legs. Such large size is
possible in the ocean, where the water supports most of the
weight of the animal. But on land, legs with an exoskeleton

thick enough to support such a load above the surface of the ground would be too heavy for much movement. Thus, terrestrial arthropods are limited, by their exoskeleton, to a relatively SMALL SIZE, which is not without its compensations if we are to judge from the success of these animals. Small size, combined with great development of the muscles, makes for active habits and easy escape from enemies. In addition, small size requires relatively little growth, and many forms develop from the egg to the sexually mature adult in a few days or weeks. Such a SHORT LIFE-CYCLE results in many generations in a year; and this means that such species may undergo *rapid evolution*, which explains, in part, the great numbers of species of terrestrial arthropods, particularly insects.

Like most invertebrates, arthropods lay large numbers of small eggs; usually the young hatch from the egg in an immature state and must feed to obtain the necessary materials for further growth and differentiation. The *larvas* of aquatic forms are free-swimming and undergo a gradual change into the adult. In the most specialized insects, the larvas (caterpillars, grubs, etc.) are so different from the adult that it is not possible to have a gradual change. The larva surrounds itself with some kind of protective material and becomes transformed into the PUPA, which is referred to as a 'quiescent' stage – and so it is, from all external appearances. Internally, however, many important changes take place, for the larval tissues break down and become reorganized through the growth of certain cells which were set aside early. After a time the sexually mature adult emerges. Such radical changes from larva to adult are known as METAMORPHOSIS ('change in form'). (See chap. 24 for examples.)

In many insects which undergo metamorphosis there is a marked DIVISION OF LABOUR AMONG THE DIFFERENT PHASES OF THE LIFE-HISTORY. In the butterfly, for example, the caterpillar has chewing jaws and feeds on leaves. It eats large quantities of food in a relatively short time and grows very rapidly. Thus, its role in the life-history is *feeding*. The pupa undergoes profound changes in structure, and its function may be said to be that of *transformation* and *differentiation*. The adult is a winged form which has no chewing jaws and can feed only by extracting nectar from flowers by means of a long sucking tube. Its ability to fly makes it important in *distribution* of the species, but perhaps its chief role is that of *reproduction*, for it does not grow and much of its food goes to produce the eggs. Moreover, some adults never feed at all but mate soon after emerging from their pupal cases; the females lay eggs and the adults die.

Specialization between different stages in the life-cycle may also involve more than one habitat. This enables one species to exploit two very different sources of energy, increasing the amount of energy available in any locality for the growth of that species. For example, the adult mosquito is a flying, terrestrial form that sucks blood, while the larva lives in fresh water and feeds on minute organic particles wafted into the mouth by special bristles.

Many of the more primitive insects have no metamorphosis or have only an INCOMPLETE METAMORPHOSIS, like that of the cockroach. The cockroach hatches from the egg as a young form which looks like a miniature adult except for some differences in general proportion and in the possession of only the beginnings of wings. (See photographs on plate 92.) It leads a life like that of the adult and grows rapidly. Naturally, an animal with a hard outer covering cannot grow indefinitely without making some kind of readjustment. And the cockroach periodically sheds or MOULTS the outermost layers of the cuticle. When the cuticle ruptures and is cast off, the animal already possesses a new 'roomier' cuticle, which has formed beneath. But until the hardening substances are laid down, the newly-shed cockroach has a light-coloured and delicate cuticle, which is elastic and stretches to accommodate the animal. From moult to moult, as growth continues, there is a gradual increase in specialization until the fully mature form is attained. Most insects do not moult after the adult stage is reached; but crustaceans, centipedes, millipedes, and arachnids do.

THE phenomenon of POLYMORPHISM, as we saw it in coelenterates, was a division of labour among the structurally differentiated subindividuals of a colony. Arthropods are not colonial in the structural sense; that is, the different individuals are not physiologically connected, as in the obelia. But many insects show polymorphism in that different members of the species are structurally specialized for the performance of different functions. They live together as co-operating members of a SOCIAL COLONY. The social insects most familiar to everyone are the highly evolved ants, bees, and wasps; but the lowly termites, relatives of the cockroaches, have one of the most interesting types of social structure. The termite *workers* are sterile; they build the nest, collect food, care for the king and queen, and raise the young. The *reproductives* are fertile and hatch as winged forms which fly out to establish new colonies. They mate; and the female, or *queen*, spends the rest of her life laying eggs. She cannot feed herself but is fed and cared for by the workers. The *soldiers* cannot feed themselves; they protect the colony from invaders. (For further details about the termites, see photographic section, plate 94.) Social life in insects has the same advantages as the social life of man. One individual need not perform all the necessary labours, the various duties being distributed among different individuals specialized for the job.

In human society the individuals are not born anatomically suited to their various occupations but become trained physically and mentally to fit their particular jobs. Among arthropod societies the individuals are *structurally adapted* from the very start, and are so specialized that sometimes they cannot even perform such an ordinary activity as feeding.

The polymorphism of social arthropods extends also to their BEHAVIOUR. The behaviour patterns of a termite, for example, are established at the outset, little or no learning being necessary for the animal to take its place in the life of the colony. This is demonstrated every time a new colony is formed. The workers which hatch from the eggs laid by the queen have never seen the nest from which the queen came; yet they construct a nest exactly like it. Such complex inherited behaviour is unlearned, or INSTINCTIVE. Not all the behaviour of social insects is instinctive. If a beehive is moved, the bees return first to the original spot; but when, afterwards, they find the new location, they 'learn' the new place.

Instinctive behaviour is superior to learned behaviour for animals, such as the insects, which live only a few days, weeks, or months, and can ill afford to spend time learning how to catch prey, eat, build a shelter, and lay eggs. Life is short, and there are many things to do. On the other hand, learned behaviour has distinct advantages for animals which live a long time and have parental care and training. The human infant is helpless and would die if left without care. Years of training are a necessary preparation for an independent life in human society. However, men can learn new kinds of behaviour and solve new kinds of problems throughout life, whereas the arthropod is more or less limited to the original set of instinctive reactions.

THROUGH the animal groups we have seen many *successive levels of structural differentiation.* Specialization within single cells, among cells, among tissues, and among organs we saw respectively in protozoans, sponges, coelenterates, and flatworms. The beginnings of segmental specialization were already apparent in annelids. Division of labour among different individuals and among the several stages in the life-history occurred in coelenterates. All of these specializations reach their extremes in the arthropods, which represent the peak of invertebrate evolution.

CHAPTER 23

The Lobster and other Arthropods

THE appendages of vertebrates are four in number; and though they show a variety of structure in adaptation to different methods of locomotion and to additional services which they may perform, such as digging or holding prey, they are primarily locomotory – with the notable exception of the fore limbs of man. The appendages of arthropods, however, are greater and more variable in number; and some of them have no locomotory function, but serve as sense organs, jaws, mating organs, or respiratory structures. Further, in contrast to the versatile limbs of vertebrates, arthropod appendages are often specialized for a single function. Thus, to describe the appendages of an arthropod is to tell almost everything about the habits of the animal: where it lives, how it moves, and how it feeds.

Primitively, arthropods had along the whole length of the body a series of simple, flattened appendages which were all alike. Each served several functions: locomotory, food collecting, respiratory, and perhaps also sensory. Such a condition is found in no living arthropod, but something very much like it is seen in the extinct trilobites (chap. 27). Among living arthropods the closest approach to this occurs in certain crustaceans, the fairy shrimps, which have a specialization of appendages on the head and a loss of appendages on the posterior region, but have on the trunk region a series of similar flattened appendages for swimming, food-collecting, and respiration. At the other extreme is an insect like the honeybee, in which every pair of appendages on the body is different. To bridge the gap between the fairy shrimp and the honeybee would be difficult if it were not that among the other arthropods we find almost every intermediate stage. The lobster, for example, has a pair of appendages on almost every segment. Those on the abdomen are simple swimming flaps which are almost all alike, but those of the head and thorax are highly diversified in structure and function. Further, in the development of the lobster embryo even the appendages of head and thorax appear first as simple, similar structures which only gradually become specialized and differentiated from each other. Thus, by tracing the development of the lobster appendages and homologizing the different parts of each, we are better able to understand how a simple, flattened swimming oar can, by gradual changes, become a chewing jaw or a sensory antenna.

THE LOBSTER

APART from minor details, the lobster is so much like its fresh-water kin, the crayfish, that the description of the lobster applies, in general, to both animals.

The body of the lobster consists of twenty-one segments (or less, if certain segments which lack segmental appendages are not counted). The first fourteen are united into a large CEPHALO-THORAX, which represents the combined head and thorax. The fusion is complete dorsally and at the sides, but the segmentation can still be recognized on the ventral surface. The ABDOMEN consists of seven distinct segments, which are clearly marked, externally.

The cuticle, secreted by the underlying epidermis, covers every part of the body, forming a jointed EXOSKELETON which is made particularly hard by an infiltration with calcium salts. The cuticle also furnishes some internal support to the cephalothorax by means of thin plates of cuticle secreted by infoldings of the epidermis. These plates increase the area for the attachment of muscles, and they protect important organs. Over the dorsal surface and sides of the cephalothorax the calcified cuticle forms a single large shield, the CARAPACE. Over the abdomen it is folded between segments, allowing for flexibility.

The LEAST MODIFIED APPENDAGES are those of the ABDOMEN. Each consists of a *basal piece* (protopodite), which bears at its

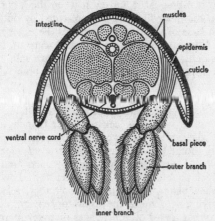

CROSS-SECTION OF ABDOMEN of the lobster.

free end an *outer branch* (exopodite) and an *inner branch* (endopodite). The numbers of joints in the three pieces may vary, but the basic plan of this two-branched (biramous) appendage is thought to be the fundamental plan of all crustacean appendages; and we find it throughout the group, both in highly specialized adults and in the simplest larvas. In the lobster the two-branched plan is obscured in the appendages of head and thorax by the presence of additional lobes or extensions on the basal piece or by the absence of the outer branch. In the lobster embryo, however, all the appendages arise as simple two-branched structures. (See figure on p. 290.)

The HEAD of the lobster is fused with the thorax, but its component segments have been determined from careful studies of the embryology of appendages and other structures, especially the nervous system. As in all arthropods, there are six segments. On the first there is a pair of COMPOUND EYES set on the ends of jointed, movable stalks. These are not serially homologous with the other appendages, since they arise in a different way. The second segment bears the FIRST ANTENNAS, sensory structures which have two filaments. (While the antennas themselves are serially homologous with the rest of the appendages, their two-branched condition is not. For in the lobster embryo the first antennas remain single until long after the other appendages have become two-branched; and even when the larva emerges from the egg, the inner filament is represented by only a small bud from the base of what finally becomes the outer filament.) The SECOND ANTENNAS, located on the third segment, have only one long filament; this is serially homologous with the inner branches of other appendages, and the outer branch is represented by a scalelike process. (From this point on, only the location and function of the appendages will be mentioned. All are paired and are serially homologous with each other; some bear extra processes or lack the outer branch, as can be seen in the diagram of the appendages.) The fourth head segment bears toothed JAWS (mandibles) for crushing the food. On the fifth and sixth segments are the FIRST and SECOND MAXILLAS, which pass food on to the mouth. The second maxilla is a thin, lobed plate and is chiefly respiratory, serving as a 'bailer' for driving water out of the respiratory cavity.

The THORAX has a pair of appendages on every segment. The first three bear the FIRST, SECOND, and THIRD MAXILLIPEDS. These are somewhat sensory but serve chiefly to handle food, mincing it first and then passing it on to the mouth. Only the third is powerful enough to do much real chewing of the food, unless it is soft. In each the basal piece bears a thin flap (epipodite), to which, on the second and third maxillipeds, is attached a gill. The flaps separate and protect the gills. The fourth thoracic segment bears the large claws or PINCHING LEGS (chelipeds), used both in offence and defence. The next four segments have each a pair of WALKING LEGS. All five pairs of legs have attached to their bases a gill-separator and a gill. The walking movements of the legs move the gills and stir up the water in the respiratory cavity under the carapace. The pinching legs are not symmetrical in lobsters over 1½ inches long. In the smallest lobsters both of them are slender and have sharp teeth; but as

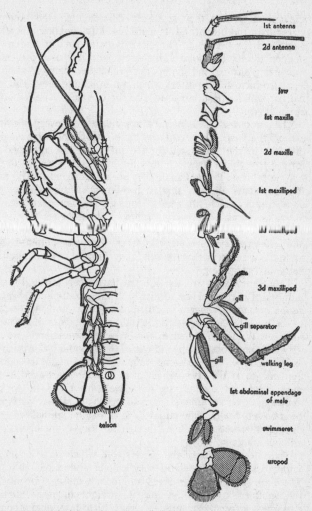

1st antenna

2d antenna

jaw

1st maxilla

2d maxilla

1st maxilliped

2d maxilliped

gill

3d maxilliped

gill

gill separator

gill

walking leg

1st abdominal appendage
of male

swimmeret

uropod

telson

APPENDAGES of the lobster show a marked division of labour. Some have extra lobes or other processes, and some lack the outer branch; but they all can be reduced to a common basic plan. The *inner branches* are stippled, the *outer branches* are shaded with diagonal lines, and the *basal piece* and its processes are left unshaded.

the animal grows, they gradually differentiate. One becomes larger than the other, and its teeth fuse into rounded tubercles; it is used for crushing. The other remains smaller and more slender, the teeth become still sharper, and it is used especially for seizing and tearing the prey. The first two pairs of walking legs also have small pincers which aid in seizing prey. The last pair of walking legs are used also for cleaning the abdominal appendages.

The ABDOMEN has a pair of appendages on every segment except the last. Those on the first abdominal segment are different in the two sexes. In the male they are modified to form a troughlike structure used for transferring sperms in the mating process. In the female they are much reduced. The next four segments all bear similar two-branched appendages, the SWIMMERETS, which function in forward swimming and in the female serve as a place of attachment for the eggs. The sixth abdominal appendages are the UROPODS, which resemble modified and enlarged swimmerets. Together with the flattened last abdominal segment, the TELSON, they form a tail-fan, used in backward swimming.

The appendages of the lobster have been stressed for a number of reasons. They furnish a striking example of *specialization among appendages* of different segments and, in the case of the large pincers, between the right and left sides of the same segment. While the flattened, two-branched swimmerets are not very different from the appendages of the hypothetical arthropod ancestral type, on the same animal we find such specialized appendages as the jaws, which have a counterpart even in the most advanced insects. In the development of the lobster appendages we see how a series of originally similar parts can become gradually differentiated into highly specialized and dissimilar structures which, though no longer analogous, are still homologous.

The internal parts of the lobster with which some of us are familiar are the large (and very edible) abdominal MUSCLES. These are segmentally arranged and include muscles for moving the swimmerets, extensor muscles for straightening the abdomen, and much larger flexor muscles, which furnish the major source of power for locomotion. For rapid movement the lobster flexes the abdomen ventrally and with such force that the whole animal shoots backwards through the water. In the cephalothorax are numerous muscles for moving the appendages and certain organs. Most of the muscles of the lobster are of the striated type, characteristic of arthropods and vertebrates. STRIATED MUSCLES contract very rapidly and are therefore well suited for moving the body and appendages. Both arthropods and vertebrates have unstriated muscles for organs such as the digestive tube and blood vessels, which undergo slow, rhythmic contractions. Lower invertebrates possess the slower, unstriated type.

The DIGESTIVE SYSTEM of the lobster consists of three main

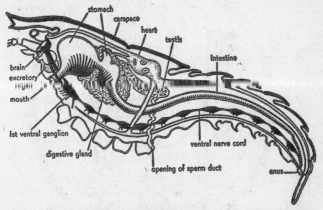

INTERNAL ANATOMY of the lobster, which, like most highly specialized segmented animals, has single large organs (or a pair of organs), instead of small local representatives in each segment. Of the systems represented here, only the nervous system is clearly segmental. Segmental blood vessels of the abdomen, omitted here, are shown in the diagram of the circulatory system. (Based partly on Herrick.)

regions, of which only the middle one has an endodermal lining. The anterior and posterior ends develop as tubular ingrowths of the ectodermal epithelium and so become lined with a cuticle which is continuous with the exoskeleton and is shed when the animal moults. Lobsters are scavengers, but they also catch live fish and dig for clams; and they have been seen to attack large gastropods, breaking off the heavy shell, piece by piece, to obtain the soft inner parts. The food is shredded by the maxillipeds and maxillas and then further crushed by the jaws before it enters the mouth. As if this were not enough, part of the stomach is specialized as a gizzard, which is lined with hard chitinous teeth and worked by numerous sets of muscles. In the stomach the food is pulverized, strained and sorted. The smallest particles are sent in a fluid stream to the large digestive glands for digestion and absorption; larger particles go in a steady current to the intestine; and the coarsest particles are returned to the grinding mechanism.

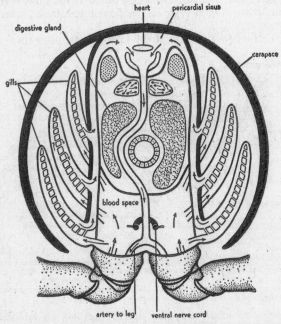

CROSS-SECTION OF THORAX of lobster to show relations of gill chambers to other organs and the path of the blood through some of the main blood channels.

The anterior portion of the stomach is large and bulbous and serves chiefly for storage. The posterior part is mainly for sorting and straining. Between the two lies the grinding region, which reduces the food to minute particles. Since these are readily digested in the tubules of the digestive glands, the work of the intestine is less important than in the earthworm. This explains how a large animal like the lobster can get along with such a short uncoiled intestine.

The extensive respiratory surface needed to supply the demands of a large and active animal like the lobster is furnished by twenty pairs of GILLS, feathery expansions of the body wall, which are filled with blood channels. The gills are attached to the bases of the legs, the membranes between the legs, and the wall of the thorax. They lie on each side of the body in a cavity enclosed by the curving sides of the carapace. Water enters the cavity under the free edges of the carapace, passes upward and forward over the gills, and is directed out anteriorly in a current maintained by the flattened plates of the second maxillas.

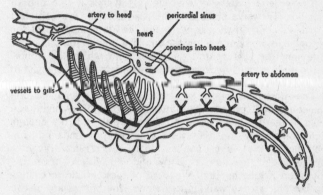

CIRCULATORY SYSTEM of the lobster, showing the main blood channels. Blood returning from the tissues goes through the gills before returning to the heart. (Modified after Gegenbaur.)

The CIRCULATORY SYSTEM is an OPEN one. The muscular heart lies dorsally in a chamber filled with blood. In the sides of the heart are three pairs of openings through which blood from the chamber enters the relaxed heart. When the heart contracts, valves prevent the blood from going out the openings; instead, it is driven into arteries which go to the tissues of the body. The smallest branches of the arteries open, not into veins, but into blood cavities in the tissues called SINUSES. Blood returning from the tissues collects in a large ventral sinus and from there enters the gills, where it gives up carbon dioxide and takes up oxygen. Then it is returned, through a number of channels, to the large pericardial sinus which surrounds the heart.

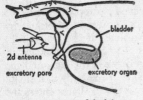

EXCRETORY ORGAN of the lobster.

The single pair of EXCRETORY ORGANS, sometimes called the *green glands* because of their greenish colour, consist each of a glandular sac and a coiled tube which opens into a muscular bladder. Wastes extracted from the blood are poured into the bladder and from there emptied to the outside through a pore at the base of the second antenna.

The general pattern of the NERVOUS SYSTEM is like that of annelids. The large brain is in the head near the eyes. From it a pair of connectives pass ventrally, one on either side of the oesophagus, and unite below the digestive tract to form a double ganglion, the first ventral ganglion, from which the double nerve cord extends backwards, enlarging into paired ganglia in almost every segment.

The most conspicuous SENSE ORGANS are the antennas and the compound eyes. As the lobster is most active at night, and even in the daytime lives at depths where there is not enough light for clear vision, the eyes are probably secondary in importance to the *sensory bristles* which are distributed all over the surface of the antennas, body, and appendages – from fifty thousand to one hundred thousand of these bristles occurring on the pincers and walking legs alone. The bristles are of two types – one sensitive to touch, and the other to chemicals. Occupying the basal segment of each first antenna is a water-filled sac which opens to the outside by a fine pore. On the floor of the sac is a ridge of sensory hairs, among which are numerous fine sand grains. As any movement of the lobster would sway the hairs or cause the sand grains to roll over them, this structure is thought to be a balancing organ.

To demonstrate the balancing function of this organ in shrimps, one investigator performed a very ingenious experiment. He obtained a shrimp that had just moulted and therefore had no sand grains in the sensory sac. He put the animal in filtered water and supplied it with iron filings. The shrimp picked up the filings and placed them in the sac. Then, when the investigator held a powerful electromagnet above the animal, it turned over on its back – apparently because the magnetic pull on the iron filings in the sac was greater than the opposing pull of gravity.

The REPRODUCTIVE SYSTEM consists of a pair of ovaries or testes, which lie in the dorsal part of the body and from which a pair of ducts leads to the external openings at the bases of the third legs in females, fifth legs in the male. The sexes can be distinguished by the position of these sex openings as well as by the structure of the first abdominal appendages. In the sex act the male deposits sperms near the female pores, and the eggs are fertilized as they emerge. They are fastened, by a sticky secretion, to the swimmerets of the female and are kept well aerated by the movements of the swimmerets.

The young lobster hatches from the egg as a free-swimming LARVA and goes through a series of changes before it comes to resemble the adult. The young crayfish, like the young of most freshwater animals, hatches as a juvenile form which is much like the adult except in size.

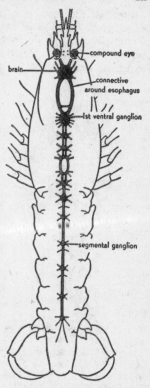

The NERVOUS SYSTEM of the lobster is much like that of the nereis, with a ring of tissue around the oesophagus and a ventral, double, ganglionated cord.

CRUSTACEANS

THE name Crustacea was originally used to designate an animal having a hard but flexible 'crust', as contrasted with one having a hard but brittle shell like that of oysters or clams. Since nearly all arthropods have a hard, flexible exoskeleton, we now use more distinctive criteria for assigning an animal to the class CRUSTACEA, of which the lobster is a member. Crustaceans may be roughly distinguished as arthropods which breathe by means of gills and have two pairs of antennas. The lobsters and crabs are giants among crustaceans; most kinds are small animals, under half an inch in length.

Although the most primitive crustaceans, such as certain brachiopods, now live in fresh water, the earliest crustaceans certainly lived in *salt water*, and the class is still predominantly marine. This is not surprising when we consider that the crustaceans as a group are the most primitive living arthropods,

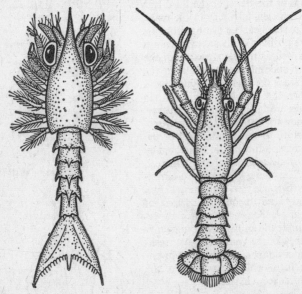

Left, the first LARVAL STAGE OF THE LOBSTER is just over ⅓ inch long. The appendages are all two-branched, similar structures. The swimmerets at this stage are only small buds, and the larva swims about at the surface by the rowing action of the flattened, fringed outer branches of the limbs. *Right,* the fourth larval stage of the lobster is about ½ inch long and resembles a miniature lobster. Like the first stage, it swims at the surface, feeding on small organisms; but forward swimming is by means of the swimmerets. The outer branches of the legs are reduced and no longer visible; the inner branches are differentiated, though right and left large claws are still similar. (After Herrick.)

and that the ocean, being the easiest place to live in, requires the fewest adjustments on the part of its inhabitants. Because of their tremendous volume, the seas provide relatively constant salt content, oxygen content, and temperature throughout the year. The salt concentration of animal tissues is much closer to that of sea water than to fresh water. Besides, sea water is buoyant, offering greater support. Crustaceans are so abundant in the ocean that they have been called 'the insects of the sea', and there is hardly any way of life in the sea not followed by some member of this diversified class.

Among the most highly modified crustaceans are the BAR-NACLES, sessile marine animals which live attached to rocks, wooden pilings, ships, and the bodies of many animals. If you are surprised to find them among the arthropods, you are like most laymen, who assume them to be molluscs because of their thick calcareous shells. Early zoologists, too, classified them with the molluscs until their true relationships were discovered through a study of their development. The young larva which hatches from the egg is free-swimming. In the possession of three pairs of appendages and in other characters it resembles the NAUPLIUS LARVA, characteristic of crustaceans. After swimming about for a time, it undergoes changes and then settles on some solid object, becoming attached by the head end. In spite of their extreme modifications, adult barnacles can be recognized as arthropods by their chitinous jointed appendages, which are two-branched, as in other crust-aceans, and are heavily fringed with bristles. The appendages are thrust out of the shell and sweep through the water like a casting net, entrap-ping small animals and organic frag-ments. With few exceptions barnacles are hermaphroditic; and, as in many other groups of animals, this is thought to be associated with their sessile life, which prevents contacts between individuals. Another curious marine crustacean closely related to the barnacles is *Sacculina*, which has a free-swimming nauplius but in the adult stage fastens on to a crab and sends rootlike processes into every part of the host's body, parasitizing

CYCLOPS, so named from the single median eye, is about 1/10 inch long and is the most familiar fresh-water copepod. Mature females usually have two groups of eggs attached to the body. Most members of the order Cope-poda are marine, but fresh-water copepods are very abundant and form an important part of the food of fishes. This food re-lationship is reversed in the case of the 'fish-lice', copepods which parasitize fish. (Modified after Herrick.)

it so completely as seriously to affect its whole physiology and arrest its growth.

The crustaceans have done almost as well in *fresh water* when we consider that this medium is a more difficult one for all groups of animals and that, compared with the ocean, which is sometimes described as a 'thick soup' because of the abundance of its animal life, fresh water hardly claims the title of 'thin lemonade'. To invade the rivers that connect directly with the ocean, a crustacean not only must become adjusted to the lowered salt content but must be able to maintain itself against the downstream current. This is not so difficult for the adults but their small and fragile larvas are easily swept downstream and back into the ocean. No doubt this has been a factor in

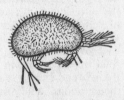

DAPHNIA is the most common fresh-water representative of the cladocerans, an order of small crustaceans (Daphnia is about 1/10 inch long, but most forms are smaller) which have a compact body often enclosed in a bivalved carapace. They swim by rapid jerks of the large two-branched second antennas. (Modified after Herrick.)

OSTRACODS are minute crustaceans, mostly fresh-water forms. The animal lives encased in a bivalved carapace. When the valves are open, as shown here, it can protrude the appendages which propel it through the water. (Modified after Turner.)

preventing some invertebrate groups from ever establishing themselves in fresh water. The ones that are successful usually suppress the free-swimming larval stages, and the young hatch as miniature adults. Further, bodies of fresh water are subject to violent fluctuations of temperature, and small ponds dry completely in the summer and freeze solid in the winter. Crustaceans have become adapted to these rigorous conditions by the development of thick-shelled eggs which resist drying and freezing.

Adaptation to *land* life is a still more difficult step. Temperature fluctuations are even more extreme, drying is a constant threat, and breathing mechanisms must be adapted to air respiration. A few crustaceans, some crabs and 'wood lice', are fairly

COPULATING EARTHWORMS usually do not leave their burrows but extend only the anterior end and mate with a neighbouring worm. The animals oppose their ventral surfaces and exchange sperms, for, though every earthworm has both male and female sex organs, it does not fertilize itself. The actual fertilization occurs later at the time of egg-laying. (Photo of living animals made at night by L. Keinigsberg)

GIANT EARTHWORMS are found in Australia. They may be located by the gurgling sounds they make as they move underground. These men are extracting a long worm from its burrow. When extended, it may be up to 12 feet long, with as many as 500 segments. The only bird known to feed on this worm is the laughing kingfisher. (Globe photo)

PULLING AN EARTHWORM OUT OF ITS BURROW is not easy, as anyone can find out by trying. The animal hangs on by inserting its bristles into the walls of the burrow. (Photo of living animal by L. Keinigsberg, Chicago)

LEECH (*Placobdella*) taken from the naked skin at the base of the hind leg of a snapping turtle. Removed to an aquarium, it attached itself to the glass by means of the two suckers, the larger of which is at the posterior end, the smaller at the mouth end. Leeches are segmented worms, but the folds of the surface are more numerous than the internal segments. A good meal lasts several months, and during this time the blood is stored in stomach pouches, which can be seen through the body wall as dark bands. The animals are hermaphroditic, having both male and female sex organs. Large specimens are 3 or 4 inches long when partly extended. The ground colour is deep olive green, with spots of brown. (Photo of living animal by S. T. Brooks. Courtesy *Nature Magazine*)

MEDICINAL LEECHES (*Hirudo medicinalis*) are still exported from Europe for the removal of black-and-blue spots, particularly around the eyes. These two were purchased in a Chicago drugstore. The leech can take three times its own weight in blood at a single meal and injects into the wound an anticoagulant. *Left*, upper surface of contracted worm. *Right*, lower surface of extended worm. (Photo by P. S. Tice)

POND LEECHES are easily collected, usually unwillingly, by wading in a pond with bare feet. In addition to the well-known 'leechlike' method of movement, leeches can swim actively by undulations of the body. They feed on worms, aquatic insect larvas, and even other leeches. When sexually mature, they take a large blood meal, which may be slowly digested over as long as a year. (Photo by Cornelia Clarke)

(Photo of living animals. Pacific Grove, California, U.S.A.)

SIPUNCULIDS are marine wormlike animals which, though unsegmented, are related to annelids. They swallow great quantities of mud or sand, from which they extract organic material. When removed from their burrows, they alternately evert and retract the anterior end of the body, which has at its tip a circle of tentacles surrounding the mouth. The one at the *left* has the anterior end retracted; the one at the *right* is fully everted. A model of a sipunculid can be seen in the mud in the picture of *Chaetopterus*.

1. ECHIUROIDS are wormlike, unsegmented animals related to annelids. *Urechis* lives in mud flats on the West Coast of North America and can be dug up during a very low tide.

2. It inhabits a U-shaped burrow, whose exact position is determined by inserting a rubber tube into one opening and blowing until water spouts from the other.

3. Following a path of the rubber tube, the mud is rapidly shovelled away until the burrow is laid open and the pink, cylindrical worm is exposed.

4. Removed to the laboratory, *Urechis* is an excellent source of eggs or sperms, which can be sucked up with a glass pipette inserted into the sexual openings.

5. Through the microscope, biologists observe the fertilization and development of the egg under normal and also under various experimental conditions.

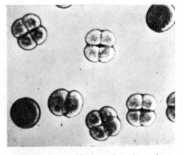

6. Highly magnified under the microscope, are one-, two-, and four-cell stages – practically indistinguishable from the early stages of most other animals.

FAIRY SHRIMPS are not really shrimps but belong to the most primitive group of crustaceans. These (*Eubranchipus*) live in fresh water; their relatives, the brine shrimps, live in salt lakes. They row themselves about, on their backs, by means of numerous, similar, flattened appendages. (Photo by Peltier. Courtesy *Nature Mag.*)

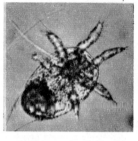

MARINE COPEPODS are minute crustaceans (1/20 inch long) which occur in countless millions in the surface waters of the ocean. They feed on microscopic plants and animals and in turn are fed on by all young fishes and many adult ones. (Photo of living animal. Pacific Grove, California, U.S.A.)

NAUPLIUS LARVA is the first stage after hatching of many freshwater and marine crustaceans. It is unsegmented and has three pairs of appendages, by which it swims about in jerks. (Photo of living animal. Pacific Grove, California, U.S.A.)

AMPHIPODS are crustaceans flattened from side to side. Some amphipods live in fresh water, but most are marine. The 'beach fleas', so-called because they are flattened and jump about, live a more or less terrestrial life on ocean beaches, feeding on plants and animals washed up by the waves. (Photo courtesy *Nature Magazine*)

LAND ISOPODS, also called 'wood live' or 'sow bugs', are among the few successful land crustaceans. They are found under logs and stones and feed on decaying vegetation. Because their delicate gill-like breathing organs (modified abdominal appendages) must be kept moist, they live mostly in damp places. The young develop in a brood pouch and emerge as young forms much like the adult except in size. (Photo by Cornelia Clarke)

AQUATIC ISOPODS ('legs all alike') live in fresh and salt water. These are *Limnoria lignorum,* marine, wood-boring forms, shown here in their burrows. Tiny animals, ½ inch long, they occur in large numbers and cause wholesale destruction of wooden pilings which support wharves. (Photo courtesy *Nature Magazine*)

BARNACLES are sessile marine crustaceans that grow mostly on rocks and on the bottoms of ships, but may live on almost any hard object in the water on which the free-swimming nauplius larva happens to settle down. This old and sluggish lobster is covered with barnacles, but healthy lobsters manage to keep free of them. (Photo of living animal by F. Schensky, Heligoland)

STALKED BARNACLES. *Right,* barnacles on the tooth of a whale. *Left,* gooseneck barnacles (*Lepas*) hanging by heavy stalks from a piece of driftwood. The two-sided shell looks like that of a mollusc, but the chitinous jointed appendages (seen protruding from some of the shells) are proof of arthropod affinities. (Photo by J. F. Pilcher, Galveston)

MODELS OF ROCK BARNACLES, *Balanus. Left,* section through a barnacle with appendages withdrawn, as when disturbed or when the tide is out. *Right,* external view of same animal with extended appendages. Barnacles have been described as animals which sit on their heads and kick food into their mouths. (Photo courtesy American Museum of Natural History)

THE FRESH-WATER SHRIMP, *Palaemonetes,* is a true shrimp, very similar to certain of its marine relatives. Actual length, $1\frac{1}{2}$ inches. Collected at Wolf Lake, Indiana, U.S.A. (Photo by P. S. Tice)

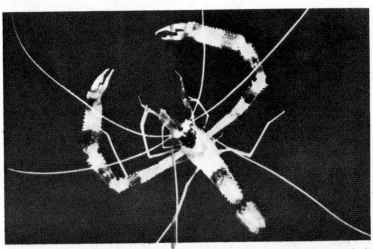

MARINE SHRIMP, *Stenopus,* which shows the general tendency of animals in tropical waters toward bizarre shapes and brilliant colours. This shrimp is white with bands of iridescent blue-green, red, orange, and purple. About $\frac{1}{2}$ natural size. (Photo of living animal, Bermuda)

CRAYFISHES live in streams and ponds and are very similar to their marine relative, the lobster. They are scavengers, feeding upon decayed organic matter, and also catch small fish. Like shrimps and lobsters, they can walk forward slowly but in escaping enemies shoot backward by suddenly contracting the powerful abdominal muscles. (Photo of living animal. Courtesy Shedd Aquarium)

THE LOBSTER, *Homarus*, is mostly dark green when alive; but when boiled, like this one and like millions of others every year, turns bright red. About half an hour after this picture was taken this lobster was reduced to an empty exoskeleton.

SWAMP CRAYFISH, *Cambarus*, beside the 'chimney' which surrounds its burrow. The burrows are 1–3 feet deep, have at the bottom a water-filled cavity, and are built in swamps and meadows, often far from a stream. (Photo by C. Clarke)

THE SPINY LOBSTER (*Panulirus*) has no large pincers and would seem to be an easy animal to approach. But the body is covered with a formidable array of spines, and the spiny antennas are large and deal vicious, tearing blows. The flesh is delicious, and the animals are eaten extensively. In the U.S.A. they occur in the warm waters off Florida and off the coast of southern California. (Photo of living animal. Courtesy Shedd Aquarium)

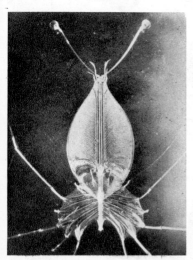

THE LARVA of the spiny lobster is a bizarre, leaflike, transparent animal with eyes at the ends of long stalks. The head and thoracic appendages are present, but the abdomen is under-developed. After several moults, the animal comes to look like the adult. (Photo by William Beebe)

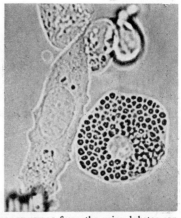

BLOOD CELLS from the spiny lobster can be kept alive for a considerable time outside the body. Two are shown highly magnified; the cell on the right was only 1/250 inch in diameter and moved about actively in ameboid fashion. The blood of larger crustaceans clots readily and is bluish from the presence of haemocyanin, a copper-protein compound which carries oxygen. Haemoglobin, found in man, occurs in some of the smaller crustacea. (Photo of living cells. Bermuda)

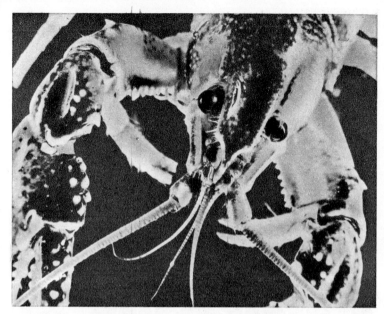

HEAD OF CRAYFISH, showing the stalked eyes and jointed antennas, the first pair short and two-branched, the second pair long and single. The two pairs of antennas are a distinguishing characteristic of crustaceans. Insects, millipedes, and centipedes have one pair; arachnids have none at all. (Photo by P. S. Tice)

THE EGGS OF THE CRAYFISH develop while attached to the swimmerets (*left*). They hatch into young (*right*) that look like miniature adults and cling for a time to the swimmerets. (Photos by Cornelia Clarke)

FIDDLER CRAB is so named from the way in which the male brandishes the huge left pincer – extending it out and then suddenly drawing it in. The 'fiddling' becomes especially vigorous in the presence of a female. The female has two small pincers of equal size. These crabs are scavengers and live in burrows on sandy ocean beaches. (Photo courtesy American Museum of Nat. Hist.)

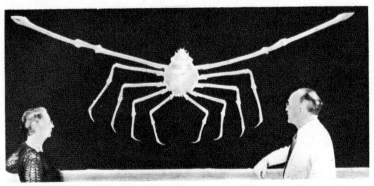

A GIANT CRAB from the waters off Japan. As in the fiddler and all true crabs, the abdomen is reduced and bent under onto the ventral surface; it cannot be seen from above. (Photo courtesy Buffalo Museum of Science)

THE HERMIT CRAB is not a true crab, for it has a long abdomen, which is soft and spirally coiled and is inserted into the empty shell of some marine gastropod. The abdominal appendages are atrophied, but one pair has hooks for holding on to the inside of the shell. The cephalothorax can be protruded, and the animal walks about carrying its shell, into which it retreats at any sign of danger. (Photo by Otho Webb, Australia)

THE KING CRAB, *Limulus,* often called the 'horseshoe crab' because of its shape, is no crab at all, but a primitive marine animal sometimes put in the same group with scorpions and spiders, which it resembles in many details. It lives along the Atlantic Coast of North America, scooping its way through the sand or mud as it hunts for the bivalves and worms, especially nereids, on which it feeds. (Photo by L. W. Brownell)

THE SCORPION is an arachnid common in the S.W. U.S.A. It hides in crevices during the day and comes out at night to hunt spiders and insects, which it catches with the large pincers, stings to death with a poison injected by the curved spine at the tip of the tail, and then sucks dry. The sting of the scorpion is painful but not dangerous to human adults, though it sometimes proves fatal to young children. (Photo by P. S. Tice)

UNDERSIDE OF LIMULUS shows walking legs and other appendages that lie under the protective hood. To the flat abdominal plates are attached the leaflike gills.

TARANTULAS are large spiders which hide during the day in the cracks of trees and under logs, stones, or debris, and at night come out to stalk their prey. They are common in the South and Southwest of America, where they reach a length of 2 inches; their bite is painful but not dangerous. Some South American tarantulas have a 7-inch span and sometimes catch small birds. Most people insist they are revolted by the long legs and hairiness, but no one on record has ever objected to these same characteristics in a Russian wolfhound. (Photo by Lee Passmore)

A GARDEN SPIDER spins a web and waits patiently until a small animal becomes entangled. Then it rushes out, seizes the prey, injects a poison and a digestive juice, and sucks up the tissue fluids. (Photo by Cornelia Clarke)

THE TRAP-DOOR SPIDER lives in a silk-lined burrow and waits, just beneath the hinged trap door that closes the burrow, for passing prey. Here the spider is pouncing on a sow bug (a small land crustacean). (Photo by Lee Passmore)

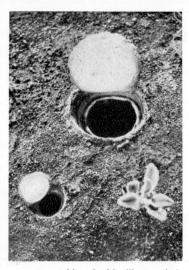

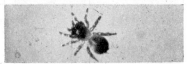

TRAP DOORS, hinged with silk, guard the entrance to the burrow of the trap-door spider. When closed, the door matches the surroundings perfectly. The size of the door is an indication of the size of its occupant.

REGENERATION of a missing leg occurs if the spider is young and growing. The adult trap-door spider, *above*, has regenerated the right second walking leg. The young spider, *below*, is a miniature of the adult.

JUST AFTER MOULTING the trap-door spider, *left,* has a white, delicate cuticle and is helpless; but in a day or so the cuticle hardens and darkens. The old cuticle, *right,* was first loosened by a moulting fluid; then it split along the sides and was shed. (Photos on this page by Lee Passmore)

BLACK-WIDOW SPIDERS (*Latrodectus mactans*) are the only really dangerous spiders in the United States, and, though common in the South and Southwest, have been reported throughout the country. The male (*left*), as in most species of spiders, is small and harmless. The female (*right*) is 1/2 inch long, black, and has a red mark, shaped like an hourglass, on the ventral side of the abdomen. The venom is very poisonous, and the bite is followed by pains and fever. Victims usually recover after two weeks, but fatalities occur.

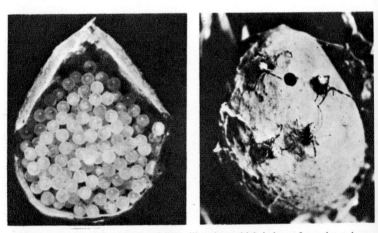

THE EGGS OF SPIDERS are enclosed in a silken bag which is hung from the web or some solid object or is carried about by the female. *Left*, opened egg sac of black widow. *Right*, the young spiders, just emerged from the egg sac, look like miniature adults. (Photos on this page by Lee Passmore)

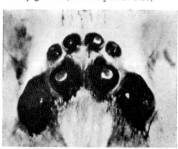

A SPIDER IS DISTINGUISHED externally from an insect by its four pairs of walking legs; a body composed of only two regions, cephalothorax (head-thorax) and abdomen; a lack of antennas; only simple eyes, usually eight in number, as in this huntsman spider; and in the possession of only two pairs of appendages about the mouth. These are the *cheliceras,* shown here beneath the eyes as broad appendages with fang-like tips turned in toward the body. and *pedipalps* – leglike appendages on either side of the cheliceras. Insects have three pairs of legs, three body regions, two antennas, two large compound eyes, very different mouth parts, and usually two pairs of wings. (Photo by Otho Webb, Australia)

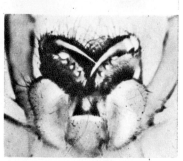

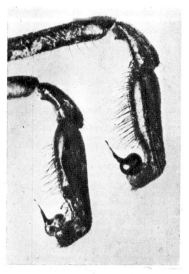

UNDERSIDE OF SPIDER'S HEAD shows fanglike cheliceras which bite prey and inject a poison through openings near the tips, Just below them are the jaws of the spider – two flat plates, each topped by a tuft of dark hairs. They are extensions from the bases of the pedipalps and can be brought together to hold prey and to squeeze it when the spider is sucking up its fluids. (Photo by P. S. Tice)

THE PEDIPALPS are chiefly sensory, but in male spiders have their tips modified to assist in copulation. The males have been seen to spin a silken net on which they deposit sperms which issue from openings on the under side of the abdomen. The sperms are picked up by the pedipalps, stored in bulbs at their tips, and later transferred to the female at the time of mating. (Photo by Lee Passmore)

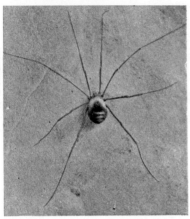

A HARVESTMAN (or 'daddy longlegs') is so called because it is seen most often at harvest time. It looks like a long-legged spider but belongs to another group of arachnids, the phalangids, because among other things, the cephalothorax and abdomen are broadly united and there are no silk glands. It feeds mostly on small insects. (Photo by Cornelia Clarke)

A MITE, as the name implies, is small. This one (shown from below) found living in an open field, is 1/4 inch long – quite sizable as mites go. Many of the parasitic types, like those that live in the oil glands and hair follicles of the human face, are only 1.50 inch long. There are four pairs of legs, as in all arachnids; the body is ovid and all in one piece. (Photo by P. S. Tice)

TICKS are large, blood-sucking mites, some of which parasitize man and his domestic animals and transmit to them serious diseases. Every year they cause millions of pounds of damage to cattle alone. The two shown at the right were kept for five years without food in the laboratory of the U.S. Public Health Service. During this long period of starvation they were able to maintain within their bodies the organisms that cause relapsing fever when transmitted to man by the bite of the tick. These ticks occur in Colorado and a few of the southwestern states of the U.S.A., but the fever they transmit it not so virulent as the one that sweeps European countries in devastating epidemics, and in Africa is a major scourge. Ticks are all parasitic, and various other mites cause diseases like mange in dogs and cattle, and itch in man; but about half the known mites are free-living on land, in fresh water, and even in the sea. (Photo from *Science Service*)

MILLIPEDES are usually found among decaying leaves and under logs in moist, shady woods or under stones in gardens. They feed mostly on decaying plants but sometimes eat living roots, becoming garden pests. Their long, cylindrical bodies consist of a head followed by a series of similar segments, almost all of which bear two pairs of legs. Millipede means 'thousand legged', and, though this is an exaggeration, some common ones do have 115 pairs. Despite the large number of legs, millipedes run slowly. They are timid creatures, avoiding enemies by hiding in dark places, often curled up into a flat spiral. (Photo of living animal)

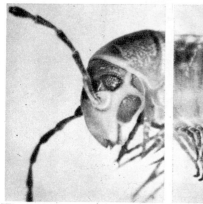

THE HEAD (*left*) of a millipede is similar in most respects to that of a centipede or an insect. It has paired antennas and chewing jaws, and a pair of accessory jaws (maxillas). There are two eyes, one of which can be seen just above the base of the antenna. It appears superficially like the compound eye of insects but is only a clump of simple eyes set close together. The posterior end of the animal (*right*) shows the terminal anus and a number of typical segments, cylindrical in shape and each with two pairs of legs. (Photo by P. S. Tice)

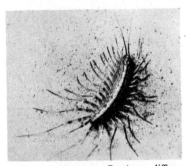

THE HOUSE CENTIPEDE, *Scutigera,* differs from other centipedes in having long, delicate legs and compound eyes. (Typical centipedes have two clumps of simple eyes.) It lives in damp places in houses, usually in the basement. It does no harm and preys on cockroaches and other insects. Slightly smaller than natural size. (Photo by Cornelia Clarke)

TYPICAL CENTIPEDES have shorter, stouter legs. The long trunk has similar segments, as in millipedes, but is flattened and has only one pair of legs to a segment. Centipede means 'hundred-legged' but *Scolopendra,* shown here, has 21 pairs of legs and the number ranges from 15 pairs as in Scutigera to 173 pairs in the geophilids. (Photo by Otho Webb, Australia)

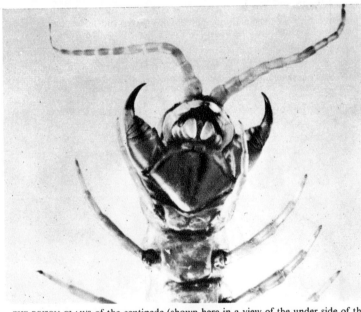

THE POISON CLAWS of the centipede (shown here in a view of the under side of the head, enlarged 3 times) are not jaws but modified appendages of the first body segment. They are curved, hollow organs, perforated at their tips, which inject a poison that rapidly paralyses prey such as insects, slugs, worms and even lizards and mice. This 6-inch Bermuda centipede inflicts a bite that may keep a man in bed with a fever for several days. Some tropical centipedes are over a foot long. (Photo by P. S. Tice)

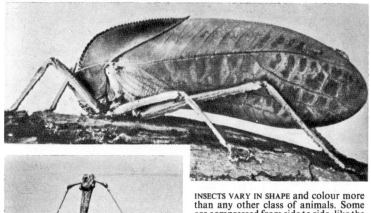

INSECTS VARY IN SHAPE and colour more than any other class of animals. Some are compressed from side to side, like the katydid, *above*; and others are flattened dorso-ventrally, like the 'leaf beetle' at the *left*. Larvas are usually cylindrical like the one at *lower left*. All the insects on this page resemble their surroundings; just how much of this is entirely accidental and how much is truly adaptive is still a hotly debated subject. In any case, it is true that the katydid and beetle, which live in dense foliage in Brazilian jungles, are green and leaflike, and the larva looks like the twigs among which it lives; while the Kallima butterfly of India, when at rest, *lower right*, looks like a dead leaf. (Photos courtesy *Nat. Mag.*)

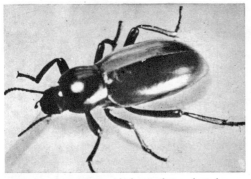

THE LEGS OF INSECTS are six in number and are borne on the thorax, the middle region of the body. All are composed of the same five joints and usually end in two curved claws. Some have sticky pads which enable the insect to walk on smooth or on vertical surfaces. The least specialized legs are simple walking legs, all three pairs much alike, as in this darkling beetle. (Photo of living animal)

THE FOOD-COLLECTING LEGS of the honeybee are also used for walking. Pollen clings to hairs on legs and body and is transferred to 'baskets' on the hind legs, here shown well loaded. (Photo by Cornelia Clarke)

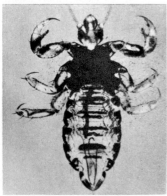

THE GRASPING LEGS of the body louse, *Pediculus humanus,* enable the animal to cling to hairs on the body of man, or more often to his clothing. With its piercing and sucking mouth parts the louse sucks the blood of its host. The danger lies not in the bite itself but in the fact that body lice transmit the deadly typhus and other fevers. The true lice, all of which suck mammalian blood, comprise the order ANOPLURA. They lack wings and metamorphosis – not primitively but by secondary loss. (Photo courtesy Army Medical Museum)

THE SWIMMING LEGS of the water-scavenger beetle are flattened and fringed with bristles. Though they live an aquatic life, these beetles have no gills but breathe by coming to the surface at intervals to obtain air, which they carry down with them as a film on the undersurface of the body. (Photo by Cornelia Clarke)

THE WINGS OF INSECTS are, typically, two pairs of membranous appendages of the thorax. They are stiffened by thickenings called 'veins', which in relatively unspecialized insects, like the damsel flies (related to dragonflies), are very fine and numerous and form a network. (Photo by Clarke)

THE THICKENED FOREWINGS of the beetles, usually called 'wing covers' form a protective armour over the body and the membranous hind wings. The elaterid beetle shown here has wing covers and hind wings spread out as if in flight. (Photo by P. S. Tice)

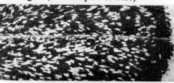

SCALY WINGS are found in moths and butterflies. The scales, shown here greatly enlarged in a close-up of a small portion of a moth's wing, are expanded, flattened bristles. (Photo by P. S. Tice)

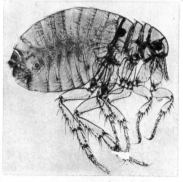

REDUCED HIND WINGS, represented by knobs called 'halteres', are found in the crane fly (shown here), mosquito, housefly, and other members of the order DIPTERA. Notice that in these relatively specialized insects the wing veins are large and few in number.

ABSENCE OF WINGS may be primitive, or the result of secondary loss as in the dog flea. The fleas (order SIPHONAPTERA) are small, wingless, insects with piercing and sucking mouth parts and complete metamorphosis. (Photo U.S. Army Med. Museum)

THE MOUTH PARTS of insects consist of the same basic parts but are modified for various methods of feeding. In the stag beetle *right,* the jaws are large and formidable, especially in the male, and are used not for chewing but for defence or holding the female during mating. These beetles are said to feed on honeydew and exudations from plants, which they gather up with their flexible, hairy labium. (Photo by P. S. Tice)

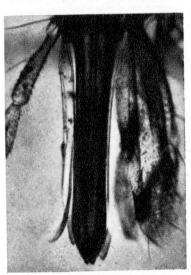

PIERCING AND SUCKING MOUTH PARTS occur in insects that feed on blood or on plant juices. The elongated parts are modified as a tube through which food is drawn up, or are stiff and sharp and used to make the wound. These last are usually fine stylets or flat blades, which may have saw edges, as in the beak, shown here, of a 'punkie' or 'sand fly.' As in moths, butterflies, and houseflies, the sucking action is provided by a muscular 'pump' in the head. (Photo courtesy *Nature Magazine*

THE SPONGING TONGUE of the housefly is the expanded, two-lobed tip of the labium. The food is sucked up through the opening at the end of the cleft between the two lobes. (Photo courtesy *Nature Magazine*)

THE SUCKING TUBE of moths and butterflies consists of the two elongated maxillas, each forming half the tube. (Photo courtesy General Biological Supply House)

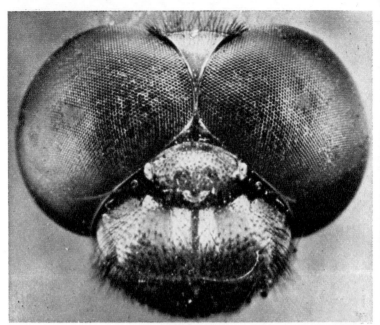

THE COMPOUND EYES of the dragonfly occupy most of the head and are composed of nearly 30,000 separate units. The antennas, which look like mere bristles, one beneath each eye, apparently play a minor role. In the housefly the eyes have only 4,000 units and in some ants there are only 50, while some nocturnal insects have no compound eyes at all and rely chiefly on their well-developed antennas. (Photo by P. S. Tice)

THE ANTENNAS of the Cecropia moth are large and branched and are important sense organs for this nocturnal animal. (Photo by P. S. Tice)

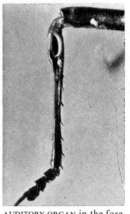

AUDITORY ORGAN in the fore-leg of a katydid. (Photo by C. Clarke)

INSECTS BREATHE by means of a system of air tubes which communicate with the outside through openings, the spiracles, located along the sides of the body, not only in adults but also in the larva, *above,* and in the pupa at *right.* Spiracles are usually provided with hairs to exclude dust and with two lips which can be brought together to close the opening, as in the spiracle *below.* (Photos of larva and pupa by Cornelia Clarke. Photo of spiracle courtesy General Biological Supply House)

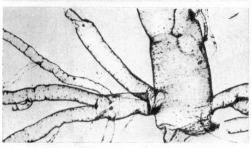

AIR TUBES BRANCH, ramifying to all parts of the body and carrying oxygen directly to almost every cell. The horizontal lines are thickenings of the cuticular lining, which help to keep the walls of the tubes from collapsing. (Photo courtesy General Biological Supply House)

AQUATIC LARVA, *Corydalus,* breathes by diffusion of gases through air tubes contained within thin-walled tracheal gills. (Photo by P. S. Tice)

THE MOST PRIMITIVE INSECTS are the bristletails, members of the order THYSANURA. They are wingless (and it is thought they never evolved any wings), have simple chewing mouth parts and no metamorphosis. The firebrat, shown here, is common in warm places, as about ovens, fireplaces, and steam pipes. Its relative, the silver-fish (which looks very much like it except for the dark markings), frequents damp, cool places, such as basements, and is commonly seen in the early morning in wash basins and bath tubs. It is a household pest, eating the starch in starched clothes and the paste or glue in book bindings. Actual size 3/8 inch. (Photo by P. S. Tice)

A COCKROACH with egg case. It frequently happens that one end of the case protrudes from the insect's body before the other end is completed. *Below left,* an egg case cut open to show the eggs. *Below right,* the newly hatched nymphs, which resemble the adult except in size and wing development. The adult, *above,* has very reduced wings, but many cockroaches have large ones capable of flight. Cockroaches have chewing mouth parts and are destructive to all sorts of organic material, scavenging any kind of food in human dwellings, and even eating holes in rayon clothing and in the covers of cloth-bound books. (Photos by Lee Passmore)

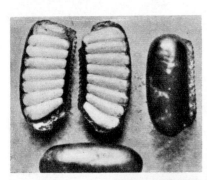

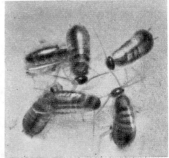

INSECT PHYSIOLOGY is a difficult subject, mostly because of the small size of the animals. Here the effect of nicotine on the heart beat of a cockroach is being studied. The movements of a fine glass thread connected to the heart by a hair are magnified through a microscope and recorded photographically. Cockroaches have also been used in growth and learning experiments. (Photo from *Science Service*)

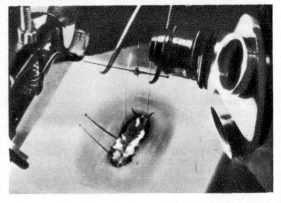

GRASSHOPPER   KATYDID

The order ORTHOPTERA includes cockroaches and the other insects on this page. Members have typically two pairs of wings (the first pair thickened and acting as covers for the second pair, which are folded fanlike when not in flight), chewing mouth parts and gradual metamorphosis, the eggs hatching into nymphs. (Photos by L. W. Brownell)

PRAYING MANTIS

Named from the way in which it holds the large forelegs while waiting quietly for its insect prey to approach. (Photo by L. Keinigsberg)

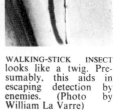

CRICKETS. *Above,* field cricket, a solitary insect that lives in fields or about houses, where it betrays its presence by chirping. (Photo by L. W. Brownell.) *Below,* tree crickets, whose high-pitched trill is one of the most conspicuous insect songs of summer nights. (Photo by Cornelia Clarke)

WALKING-STICK INSECT looks like a twig. Presumably, this aids in escaping detection by enemies. (Photo by William La Varre)

TERMITES belong to an order of highly socialized insects, the ISOPTERA, which live in colonies. They have chewing mouth parts and gradual metamorphosis. There are several castes, each consisting of both male and female individuals. *Left,* worker; *middle,* soldier; *right,* winged reproductive. These latter leave the colony, mate, and start a new colony. (Photos of living *Kalotermes,* about 5 times natural size, from Florida, U.S.A. Animals loaned by V. Dropkin)

LARGEST TERMITE QUEEN known (*Macrotermes natalensis*); actual size. The queen lays eggs at the rate of one a second, without cessation, during a lifetime of about 30 years. Head and thorax are small; only the abdomen is greatly enlarged. (Specimen loaned by A. E. Emerson)

MODEL OF ROYAL CHAMBER, showing the huge queen in the centre surrounded by a few of her millions of workers and soldiers, all of them her offspring. This species (*Constrictotermes*) has peculiar soldiers; instead of jaws, they have a projection from the head which serves as a squirt gun, through which they project a sticky secretion that entangles the enemy, especially ants. The king, who is constantly present near the queen, and like her is fed and cleaned by the workers, is seen at the lower left. (Photo courtesy Buffalo Museum of Science)

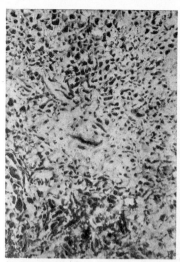

TERMITES EAT WOOD, excavating tunnels and chambers. At the centre is the large royal chamber in which the queen lives. (Photo courtesy A. E. Emerson)

NEST, or termitarium of an Australian species, showing the large size reached by some nests. (Photo by H. O. Lang, courtesy A. E. Emerson)

Above, MUSHROOM-SHAPED NEST of *Entermes,* a form which lives in tropical rain forests. The finger-like projections of the caplike roof effectively shed rain. (Photo by H. O. Lang, courtesy A. E. Emerson)

Upper right, DAMAGE BY TERMITES (*Reticulotermes flavipes*) to a book in a library at Van Buren, Arkansas, U.S.A. (Photo courtesy U.S. Bureau of Entomology)

Right, DAMAGE TO WOODEN STRUCTURES in the Middle West of the U.S.A. is trivial compared to that in the South or in tropical countries, but apparently sufficient to make termite extermination a profitable business

THE DOBSON FLY, *Corydalus* (along with other insects on this page), belongs to the order NEUROPTERA, characterized by 4 wings with many veins and cross-veins, biting mouth parts, and complete metamorphosis. (Photo courtesy Amer. Mus. Nat. Hist.)

THE DOBSON FLY LARVA, known as the 'hellgrammite' to fishermen who use it as bait, lives in rapid streams, under stones, and catches larvas and naiads of other insects. It breathes by tracheal gills. After three years it goes on land to pupate. (Photo by P. S. Tice)

LACEWING FLY. Larva, *lower left,* known as 'aphis lion', lives on vegetation and sucks the juices of plant lice and other small insects. Pupa, *upper left,* is emerging from its cocoon. (Photo by Cornelia Clarke)

ANT LION. *Upper* left, larva which digs a pit and lies buried with only the jaws protruding, ready to seize any ant or small insect that falls in. The pupa, *lower right,* is shown in opened pupal case. (About twice nat. size)

MAYFLIES are delicate animals which emerge from streams, ponds and lakes in thousands during the summer. They belong to the order EPHEMERIDA; and, as this name suggests, they live but a single day, during which time they moult twice, mate, and the female lays eggs in the water. Since the adults lack fully developed mouth parts, they never feed. The metaphorphosis is incomplete, the egg hatching directly into an aquatic naiad, which feeds on plant material. Some naiads require one to three years to get ready for their brief adult life. (Photo courtesy *Nature Magazine*)

1. THE DRAGONFLY NAIAD lives at the bottom of ponds, lakes, and streams, where it preys on small animals. Notice the developing wings and the large compound eyes. (Photo by L. W. Brownell)

2. WHEN THE NAIAD MATURES, it crawls up a plant that extends above the water. The skin splits down the head and thorax, and the adult emerges. (This photo and the two below, courtesy *Nature Magazine*)

3. THE NEWLY EMERGED ADULT has soft, limp wings which gradually expand as blood is pumped into them. When hardened (*right*), they make the dragonfly one of the fastest of all flying insects.

4. THE ADULT DRAGONFLY catches small flying insects. It belongs to the order ODONATA, characterized by four membranous, finely netted wings, chewing mouth parts and incomplete metamorphosis.

A BUG to the layman is almost any kind of insect; but to the entomologist, only the members of the order HEMIPTERA are true bugs. Hemiptera means 'half-wings' and refers to the fact that the basal half of the front wings is thickened, the posterior half membranous. The second pair of wings is membranous and folded beneath the first. The mouth parts are modified for piercing and sucking, and the metamorphosis is gradual. The southern green-plant bug, *left*, sucks the sap of garden vegetables. (Photo courtesy U. S. Bur. Entomol.) *Right*, wingless stink-bug nymphs and eggs. Stink bugs emit a fetid odour which remains on anything they visit, and this explains the bad taste of occasional berries in a box of otherwise good ones. (Photo by Cornelia Clarke)

GIANT WATER BUG, also called the 'electric light bug', is one of the largest insects, with a length of 4 inches in some species. It lives in water, swimming rapidly with its flattened hind legs, and feeding on fish, snails, and insects. It catches prey with its modified front legs and kills them with its piercing and sucking beak. The mating occurs under water, the larger female cementing the eggs all over the back of the male, *left*. The nymphs hatch, *right*, and are said to indulge in cannibalism, the older eating the younger. (Photos by Lee Passmore)

TREE HOPPERS live on vegetation, feeding on plant juices. The thorax is prolonged backward over the abdomen and often has a grotesque shape. *Left,* thorn tree hopper has the thorax prolonged into a single enormous horn; and when the animals are resting on a thorny shrub, it is difficult to distinguish them from the plant. *Above,* buffalo tree hopper, so-called from the hump and two small horns. (Photos courtesy *Nature Magazine.*) The order HOMOPTERA includes the insects on this page and others such as the cicada. Members typically have four wings (which usually are held sloping at the sides of the body when at rest), piercing and sucking mouth parts, and gradual metamorphosis.

PLANT LICE, or aphids, are tiny, and usually green, pear-shaped insects that infest almost every kind of plant. They insert their beaks into stems and leaves and suck the juices. The life-cycle is very complex and involves both asexually and sexually produced offspring and both winged and wingless forms *Above,* wingless female; *below,* winged female of pea aphid. (Photo courtesy U.S. Bureau Entomology)

MEALY BUGS, so-called because they are covered with a powdery excretion, are common pests on trees and in greenhouses. The males are usually winged, the females usually wingless and degenerate. *Pseudococcus citri,* shown here, is a serious pest of orange trees in the Southern States of America. (Photo by Cornelia Clarke)

1. THE CICADA is one of the best-known homopterans, partly because of its large size and the loud shrill song of the male, but also because certain species have interesting breeding habits which have long attracted attention. The periodical cicada, *Tibicina septendecim,* is popularly known as the 'seventeen-year locust'; but this is a misnomer, as locusts are properly migratory grasshoppers. Shown here is a female laying eggs in slits which she has made in a twig. (Photo by J. C. Tobias)

2. THE EGGS, imbedded in a twig, hatch in six weeks into nymphs, which fall to the ground and bury themselves. They suck the juices of the roots of forest and fruit trees. (Photo by Cornelia Clarke)

4. ADULT EMERGES through a slit down the back of the nymphal covering. (Photo by P. Knight, courtesy *Nature Magazine*)

3. THE NYMPHS EMERGE from the ground after seventeen years and crawl up on tree trunks. Since a whole brood emerges at one time, the empty nymphal coverings may later be seen by the thousands, clinging to the bark of trees. (Photo by C. Clarke)

5. NEWLY EMERGED ADULT with wings partly expanded. Adults mate and soon die. (Photo by Knight, courtesy *Nature Mag.*)

THE JAPANESE BEETLE is typical of the order COLEOPTERA ('sheath wings'), members of which have the forewings thickened as wing covers that protect the membranous hind wings folded beneath them. They have chewing mouth parts and complete metamorphosis. *Middle,* the larva of the Japanese beetle, called a grub, lives in the ground, feeding on the roots of grasses. *Right,* pupa. Japanese bettles do untold damage to fruit trees. (Photo courtesy U.S. Bureau of Entomology)

THE HAZEL-NUT WEEVIL is a member of the snout beetles, which have the head prolonged into a snout, provided at its tip with a pair of jaws with which it bores into plants. (Photo courtesy *Nature Magazine*)

(Photo courtesy *Nature Magazine*)

HERCULES BEETLES show the striking difference that may occur between the two sexes of the same species of insect. The male has a large horn on the head and another longer one on the thorax. The female is smaller and has no horns. The name is derived from the large size of the beetles, which are shown here about lifesize. Some of them are even larger.

THE RHINOCEROS BEETLE, a voracious herbivore, excavates burrows in palms. The three horns are on the thorax, which overhangs the small head. In spite of their heavy chitinous exoskeleton, beetles are good flyers. On the whole, the group is very successful; almost 20,000 species are known in the United States alone. (Photo by P. S. Tice)

WHIRLIGIG BEETLES swim on the surface of quiet ponds, feeding on small insects. The eyes are divided, the upper half for seeing in air and the lower half for seeing in water. (Photo by Cornelia Clarke)

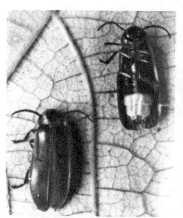

FIREFLIES are not flies but beetles; the wingless females and the larvas are called 'glow worms.' On the lower side of the abdomen are the light organs which flash intermittently. That flashes enable the males to find their mates is shown by the fact that a female in a perforated opaque box does not attract males while one in a corked glass bottle does. (Photo by C. Clarke) *Right,* photo made with the light of a firefly whose abdomen can be discerned toward the right. About 3 times natural size. (For further details on animal light see text)

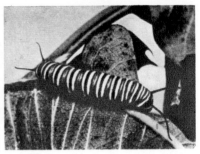

BUTTERFLIES and moths are members of the order LEPIDOPTERA ('scaly wings'), characterized by four large wings covered with overlapping scales, sucking mouth parts, and complete metamorphosis. The pictures on this page show the development of a typical member, the Monarch butterfly. (Photos by Cornelia Clarke.) *Left*, the egg (shown greatly enlarged) is laid on milkweed plants. *Right*, the larva is a caterpillar with chewing jaws; it creeps about on plants feeding almost continuously.

THE LARVA CHANGES INTO A PUPA, hanging from a leaf. From *left to right* the larval covering is shed, revealing the pupa already formed beneath.

Above, THE PUPA (called a 'chrysalis') appears to be 'resting'; but inside, the whole structure of the insect is being reorganized. *Right*, THE NEWLY EMERGED ADULT clings to the old chrysalis covering while the body and wings expand and dry.

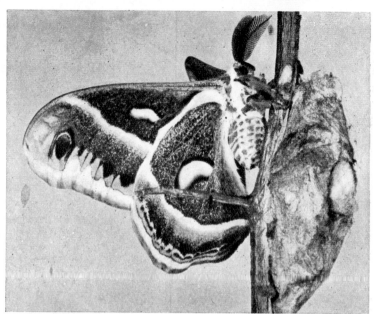

THE CECROPIA MOTH, with wings fully expanded, rests on a twig which bears the silken cocoon from which it has emerged. Moths are roughly distinguished from butterflies by their nocturnal habits, threadlike or feathery antennas, and their manner of holding the wings when at rest; the Cecropia is exceptional in holding the wings more like a butterfly; the tiger moth, below, has a more typical position. Butterflies are active during the day, have knobbed antennas, and hold the wings vertically when at rest.

COCOON of a Cecropia moth, cut open. The outlines of the antennas and wings can be seen through the pupal covering. (Photo by L. Keinigsberg)

THE TIGER MOTH rests with the wings held roof-like over the body, the characteristic position of most moths. (Photo by L. W. Brownell)

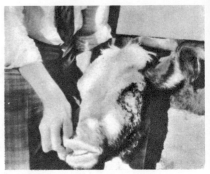

FLIES comprise the order DIPTERA ('two wings'), which also includes the mosquitoes, gnats, and midges. All have two wings (the hind wings being represented only by knobs); sucking, or piercing and sucking, mouth parts; and complete metamorphosis. They are among the most obnoxious of all insect pests. Houseflies spread typhoid fever, certain mosquitoes carry malaria and yellow fever, and tsetse flies transmit African sleeping sickness. *Chrysomyia, left,* lays its eggs in sores about the eyes of cattle, *right.* The eggs, shown enlarged *below left,* hatch into larvas, *below centre,* which are known as screw worms and feed on the tissues, causing extensive injury. Finally, they pupate, *below right.* (Photos courtesy U.S. Bureau of Entomology)

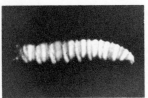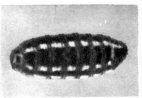

YELLOW-FEVER MOSQUITO (*Aëdes aegypti*). The male, *left,* feeds on plant juices. The female sucks mammalian blood and lays eggs in the water. The larva, *centre,* called a 'wriggler', breathes air through a spiracle at the posterior end. The pupa, *right,* breathes through two tubes that look like horns. (Photo U.S. Bur. Entomol.)

ANT workers are wingless, sterile females. Only the reproductives are winged. Ants belong, with bees, wasps, and others to the order HYMENOPTERA characterized by four wings with very few veins, mouth parts adapted for chewing or for chewing and sucking, and complete metamorphosis. Ants are scavengers, taking to their nests any kind of organic matter; some species attack living animals as well. The two shown here are tending their 'cows' (aphids), from which they obtain honeydew, a sweet exudation. (Photo by Cornelia Clarke)

ANT LARVAS, PUPAS, AND NEWLY EMERGED ADULTS are cared for by workers, which provide protection and feed the larvas. Ants live in social colonies and are the most numerous of all terrestrial animals of comparable size. (Photo by Cornelia Clarke)

HONEYBEE WORKER with pollen grains clinging to the long, branched hairs on body and legs. As with ants, the worker is a sterile female; but unlike ants, bee workers are winged. The elongated mouth parts are modified for sucking nectar; there are also chewing jaws for manipulating wax. (Photo by Cornelia Clarke)

A SWARM OF HONEYBEES includes workers, drones (males), and queen. All individuals of a colony are offspring of the queen, who may lay over a million eggs during her lifetime. At any one time, a hive may contain from 50,000 to 80,000 bees almost all workers. (Photo courtesy U.S. Bureau Entomology)

 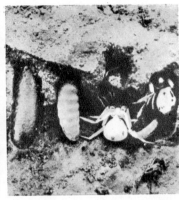

THE MUD-DAUBER WASP is solitary; but there are social wasps, just as there are both solitary and social bees. The mud dauber builds a few cells of moist earth, fills them with live, paralyzed spiders, and lays an egg in each. *Right,* the nest cut open to expose four cells containing, *from left to right,* a pupa, a larva beginning to pupate, and two cells filled with spiders, one of which has a small larva clinging to its leg. (Photos by Lee Passmore)

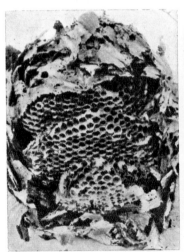 

HORNET'S NEST, cut open to show the cells in which are reared the developing young. The whole structure is made of bits of weather-worn or decayed wood chewed up with saliva to form a kind of paper. Hornets have a social organization much like that of bees; the young are fed on insects brought to them by the winged workers. (Photo by Cornelia Clarke)

OAK APPLES are growths induced on oak trees by the larva of a hymenopteran known as a 'gall wasp'. The species of parasite can be determined by the type of growth induced. The larva feeds on the green tissues, and the plant responds by overgrowth. After pupation the insects emerge, mate, and lay eggs on new oak leaves. (Photo by L. W. Brownell)

A GIANT WASP is *Pepsis formosa*, the well-known 'tarantula hawk' of the South-western States of the U.S.A. The wings show several features characteristic of the wings of Hymenoptera. The venation is reduced; and the hind wings, which are smaller than the front pair, have on their anterior margin a row of hooks which fit into a groove on the posterior margin of the forewings. This holds the wings together in flying. Wasps have a long slender 'waist' or stalk joining the abdomen to the thorax. (Photos on this page by Lee Passmore)

MANY WASPS PROVISION THEIR NESTS WITH SPIDERS; and so it is not surprising that *Pepsis*, being a very large wasp, can capture tarantulas, paralyzing them with a poison injected by the sting on the end of the abdomen, *upper right*. Sometimes the wasp loses the struggle and is eaten by the tarantula; but when Pepsis is successful, she places the tarantula in a burrow in the ground, lays her egg on its body, and closes the burrow. The larva that hatches feeds on the tarantula. Then it pupates; and when the adult emerges, it digs its way out.

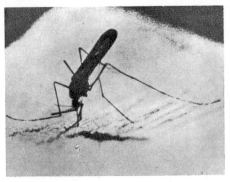

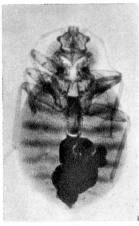

INSECTS SPREAD HUMAN DISEASES such as malaria, yellow fever, elephantiasis, African sleeping sickness, typhus, and many others; they are thus indirectly responsible for more deaths than any other agency, including war. Means of control are well known in most cases but are not applied on a wide enough scale because, among other things, men spend too much time fighting each other instead of their real enemies, the insects. Shown here is the malaria mosquito, *Anopheles,* sucking blood by inserting its piercing mouth parts through the skin. It can be recognized by the oblique position in which it holds the body while biting. (Photo from *Science Service*)

BED BUGS are flat, wingless hemipterans that live in human dwellings. They hide during the day in cracks in the furniture or floors and emerge at night to suck blood from their sleeping hosts. (Photo courtesy U.S. Army Medical Museum)

BOTFLY LARVAS attached to the lining of the stomach of a horse. The eggs are laid on the skin, licked off, and swallowed by the horse. In the stomach they hatch into larvas, which attach themselves and, when present in large numbers, cause indigestion. Finally, the larvas pass out with the faeces, pupate, and emerge as adult flies. (Photo courtesy U.S. Bureau of Entomology)

INSECTS PARASITIZE INSECTS, and so aid in the control of many insect pests. This sphynx-moth caterpillar is covered with cocoons of a braconid fly (hymenopteran), which have developed from eggs laid, by means of a long ovipositor, beneath the skin of the host. The larvas fed on the tissues of the caterpillar, emerged through the skin, spun their cocuuns, and pupated. The exhausted caterpillar will soon die. (Photo by C. Clarke)

INSECTS DESTROY MAN'S PLANT FOOD to the extent of hundreds of millions of pounds' worth every year. *Left,* a field of corn; *right,* the same field after an attack by migratory grasshoppers. (Photo U.S. Bureau of Entomology)

GRAPEFRUIT CROP DESTROYED in Florida, U.S.A., by larvas of the Mediterranean fruit fly, which hatch from eggs laid in the fruit. (Photo courtesy U.S. Bur. Entomol.)

COTTON BOLL WEEVIL AT WORK. One of the most destructive insects, this snout beetle attacks only cotton. (Photo courtesy U.S. Bureau of Entomology)

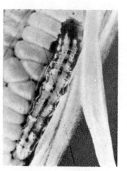

CLOTHES MOTHS AND LARVAS. Only the larvas eat woollen fabrics. (Photo courtesy U.S. Bureau of Entomology)

DAMAGE TO TREES caused by burrowing activities of bark beetles. (Photo courtesy U.S. Bureau of Entomology)

CORN-EAR WORM is a caterpillar that feeds on the tips of sweet corn. (Photo by P. S. Tice)

INSECTS ARE BENEFICIAL TO MAN in many ways. The bumble bee (shown here) and many other insects pollinate flowers whole collecting nectar for food. Without this service our orchards would produce no fruit and many crops could not be grown. Unfortunately, in warring on harmful insects we often kill useful ones; the sprays used to kill pests of fruit trees also poison the honeybees and other nectar-gatherers. Honeybees are about fifty times as valueable for cross-pollination in orchards and fields as they are for the honey they produce. (Photo U.S. Bureau of Entomology)

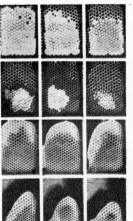

SILK produced by larvas of the silkworm moth is still a leading fabric and not yet replaced by synthetic substitutes. Silk is obtained by unwinding, either by hand or machine, the single long strand, often half a mile long, that forms the cocoon of the moth. It takes about 5,000 cocoons to make a silk kimono. (Photo by Clarke)

HONEY supplied by honeybees is our best-known insect food product. One pound of honey represents 20,000 bee trips from flowers to hive. In many parts of the world it is the insects themselves that are eaten. American Indians eat grasshoppers, crickets, and ants; many races eat caterpillars and grubs; and Africans eat termites and even fly maggots. This seems repulsive to us, but grasshoppers are clean and eat grain, while we eat their relatives the lobsters, that live as scavengers on the sea bottom. *Right,* toasted caterpillars of the Pandora moth are considered a delicacy in some parts of Mexico. (Photo courtesy *Nature Magazine*)

MOST STARFISHES LIVE ON ROCKY COASTS, where the hard substratum furnishes a place of attachment for the suckers of the numerous tube feet with which these animals pull themselves slowly over the rocks. When the tide is out starfishes can be seen clinging to the exposed algae-covered rocks, like the one in the picture *below*, or in shallow tide pools. Some starfishes can live on sandy bottoms, and these may have suckerless tube feet which are used like rows of little legs to walk over the sand. Many whole groups of invertebrates are confined almost entirely to rocky shores, especially the sessile types like sponges, sea anemones, barnacles, tube worms with calcareous tubes, and sessile bivalves. Creeping animals like chitons and most snails live where rocks afford a hard surface and a place of attachment for the algae on which they feed. (Pacific Grove, California, U.S.A.)

Above. STARFISH ABOUT TO EAT A MUSSEL (bivalve). The tube feet, which are applied to the glass of the aquarium and to the mollusc, will soon pull the two valves apart as the exhausted bivalve yields. (Photo of living animal by W. K. Fisher, Pacific Grove, California, U.S.A.)

Left. STARFISH MOPS are used to remove starfishes from oyster beds, where they cause great damage. The mops are dragged over the sea bottom, and the animals are caught when the small pincers (pedicellarias) take hold of the threads. At one time starfishes collected in this way were chopped up and thrown back. Since they regenerate easily, this method increased their numbers. (International News Photo)

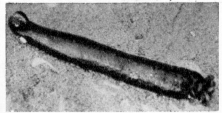

STARFISHES REGENERATE easily. This one-armed starfish is regenerating four new arms, which appear as buds growing from the disc (at the end to the right). (Photo by Otho Webb, Australia)

114

TEN-ARMED STARFISHES are less common than the typical five-armed kinds, but ones with from four to twenty-five rays are known. Notice the barnacles that cover the rock. (Photo courtesy *Nature Magazine*)

SIX-ARMED STARFISH. The larvas of some starfishes, like those of *Leptasterias,* shown here, have no free-swimming stage but are sheltered in a brood chamber on the lower side of the female and emerge as little starfishes. (Photo by W. K. Fisher, Pacific Grove, California, U.S.A.)

A STARFISH RIGHTS ITSELF (*top to bottom*) by bending its stiff arms and pulling with the tube feet. (Photo of living animal. Pacific Grove, California, U.S.A.)

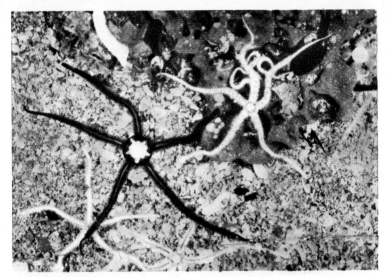

SERPENT STARS crawl over the ocean floor by agile movements of their arms, which are also used to entrap small animals like worms and crustaceans. Most species are five-armed, but ones with six, seven or eight arms are known, and some have branched arms. About $\frac{1}{3}$ actual size. (Photo by D. P. Wilson, Plymouth)

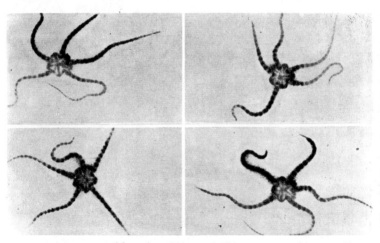

SERPENT STARS are named from the writhing snakelike movements of the arms. They are also called 'brittle stars' because of the ease with which the arms come off when seized by man or other enemies, but regeneration is rapid. About $\frac{2}{3}$ natural size. (Photos of same animal taken in rapid succession. Pacific Grove, California, U.S.A.)

UNDERSIDE OF SERPENT STAR to show central mouth from which radiate five arms. The tube feet lack suckers and ampullas and function only as tactile sense organs and for respiration. (Photo by P. S. Tice)

SEA URCHINS in a tide pool among some large anemones. Urchins live mostly near rocky shores, feeding on algae and decaying materials, and serving as one of the many kinds of scavengers that keep the sea bottom from becoming foul. In some Mediterranean countries urchins are collected by the hundreds of thousands for their sex organs, which are considered good eating. (Photo by W. K. Fisher, Pacific Grove, California, U.S.A.)

THE TUBE FEET of sea urchins end in suckers and are long and slender, extending beyond the spines. Both are used in locomotion, the tube feet being moved by changes in pressure in the water vascular system, the spines by muscles. About $\frac{1}{2}$ natural size. (Photo of living animal. Pacific Grove, California, U.S.A.)

THE TEST or skeleton of a sea urchin is globular and made up of fused calcareous plates. The spines are removed, but their positions are indicated by the round protuberances on which they were pivoted. The five double rows of holes are openings through which the tube feet protruded.

 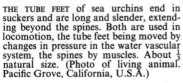

1. Unfertilized egg. 2. Fertilized egg. 3. Two-celled stage. 4. Two cells, each divide.

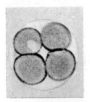   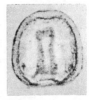

5. Four-celled stage. 6. Eight-celled stage. 7. Blastula, hollow ball. 8. Gastrula, 2 layers.

THE DEVELOPMENT OF A SEA URCHIN is representative of animal development, and equivalent stages occur in all animals from the hydra on. As the embryo divides, the cells become progressively smaller, until in the blastula stage, at the time of hatching from the egg membrane, they are barely distinguishable at this magnification. (Actual size of unfertilized egg, 1/150 inch.) (Photomicrographs by D. P. Wilson, Plymouth)

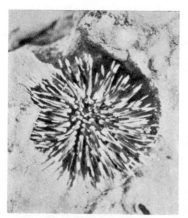

SPINES OF SEA URCHINS are used in locomotion and offer protection from marauding animals which might otherwise prey on urchins. The spines may be relatively short and stout, as in the short-spined urchin at the *left*, which sits in a hole in the rock made by constant movements of the spines; or they may be longer, very sharp, and poisonous, as in the needle-spined urchin at the *right*. (Photos by Otho Webb, Australia)

ANOTHER VARIATION IN SPINES is shown by the slate-pencil sea urchin. (Photo by Otho Webb, Australia)

SAND DOLLARS belong to the same class as sea urchins but are flattened and covered with short spines. They move by means of the spines and small tube feet on both surfaces of the body. The large tube feet which protrude from the five double rows of holes on the upper surface are respiratory. The animals swallow sand and digest the organic material contained in it. *Above,* living sand dollars. *Below,* the dried tests of two sand dollars arranged in positions corresponding to the pair above. The left ones show the upper surface; the right ones the lower surface with the mouth opening in the centre and the anus near the margin. (Pacific Grove, California, U.S.A.)

SEA CUCUMBERS are fleshy echinoderms in which the usual spines are reduced to minute ossicles imbedded in the skin. When attacked, some throw out their viscera, leaving them for the enemy, meanwhile escaping and regenerating a new set. Others throw out slime threads which entangle the enemy. The animals shown here (in a tide pool off the Australian coast) creep about on the ocean floor by muscular movements of the body wall. The tube feet are little used. The animals swallow sand or mud and digest the organic material. They are collected in great quantities and, when dried, are known as 'trepang', or '*bêche-de-mer*', which is then shipped to China to be used in making a soup. (Photo by Otho Webb, Australia)

THE SEA CUCUMBER in the foreground at left shows three of the five rows of tube feet, by which the animal attaches to rocks. The finely branched tentacles around the mouth are slimy and trap small animals. The sea cucumber at the right has the tube feet contracted. (In the background are two starfishes, one with 5 arms broadly joined to the disk, the other with 8 arms.) (Photo of living animals by F. Schensky, Heligoland)

SEA LILY, or crinoid, a stalked echinoderm, brought up from the sea bottom by a dredge. Only a small portion of the stalk is shown. When undisturbed, the branched arms are widely spread, and ciliated grooves on their upper surfaces sweep small organisms into the central mouth. Being deep-water forms, stalked sessile crinoids are seldom seen. More familiar are the feather stars of shallow water. These have a young stalked stage but later break from the stalk and swim about by waving the arms. The tube feet of crinoids are chiefly respiratory, not locomotory. The skeleton is of calcareous plates; there are no spines. (Photo by A. H. Clark. From *Science Service*)

FORAMINIFERS of the past dwarf the largest modern forms, which, though only the size of a pinhead, are giants among living protozoa. *Nummulites*, shells of which are shown about natural size, flourished on the sea bottom in the early Tertiary and formed great beds of limestone, now exposed in the Alps and northern Africa. The pyramids of Gizeh, near Cairo, are built of nummulitic limestone.

SMALL CUP SPONGE *Palaeomanon*, from the rocks laid down in the seas of middle Silurian times. (About ½ actual size)

LARGE SILICIOUS SPONGE, *Hydnoceras*, lived on the bottom of a Devonian sea in what is now New York State. (About ⅓ actual size)

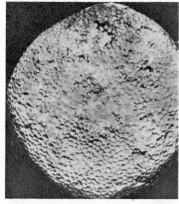

TETRACORAL, *Columnaria*, a colonial coral with internal partitions in multiples of 4. Before becoming extinct at the end of the Paleozoic, the tetracorals probably gave rise to the hexacorals, with partitions in multiples of 6, to which belong modern sea anemones and reef corals. (About ⅓ actual size)

GRAPTOLITE fossils are carbonized films left by extinct colonial coelenterates, *Diplograptus*, from the Ordovician, probably drifted by means of a central float, from which radiated numerous stems, each with a double row of horny cups occupied by the members of the colony. (About ½ actual size)

THE EARLIEST BRACHIOPODS had thin horny shells which were unhinged like those of *Lingula*, at the *left*. After the Cambrian these unhinged types were outnumbered by the hinged brachiopods with calcareous shells like *Terebratula, right*, and *Spirifer, below*. (Photo of Lingula by Mrs C. L. Fenton)

BRACHIOPOD SHELLS, when very numerous, form solid beds of limestone called 'lamp-shell coquina'. This piece of Ordovician coquina contains *Dalmanella* shells. Notice the small hole in one of the shells, probably the result of an attack by a carnivorous gastropod. (Photo by Mrs. C. L. Fenton)

CYSTOID, *Pleurocystis,* from the Ordovician. Note the broken bases of the two arms. The most primitive of echinoderms, cystoids became extinct at the end of the Paleozoic. (About ⅔ actual size)

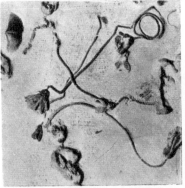

STALKED SESSILE CRINOIDS, *above,* were the dominant members of their class during the Paleozoic; about ¼ actual size. Sessile crinoids still live in deep waters, but the free-swimming types are much more common now. *Right,* crown and part of stem of a crinoid, *Dizygocrinus,* from the Mississippian. (About ⅔ natural size)

RESTORATION OF A MIDDLE-SILURIAN SEA BOTTOM, based on animals found in the Irondequoit limestone and the Rochester shale of New York. In foreground at left are two stalked crinoids. Its relatives, the cystoids, are on the reef at the right, which has been built up mainly by bryozoa. On the bottom are several kinds of trilobites, a straight-shelled nautiloid, some brachiopods, and some branching bryozoan colonies. (Courtesy Buffalo Museum of Science)

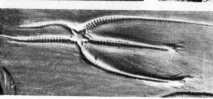

Upper left, SEA-URCHIN test, *Cidaris*, slightly enlarged, from the Jurassic. *Above*, STARFISH, *Petraster*, from the Ordovician; actual size, *Left*, SERPENT STAR, *Furcaster*, from the Devonian; About ½ actual size.

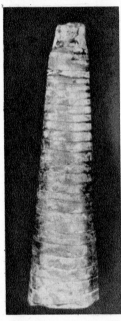

SHELLED CEPHALOPODS started out with straight chambered shells shaped like a cone, as in *Orthoceras* (about ½ actual size) at *left*. These finally reached lengths approaching 20 feet. Some of them gradually evolved curved shells, as in *Jolietoceras* (slightly less than ½ actual size), *above right*. Both nautiloids and ammonoids finally developed coiled shells, as in the ammonite, *Coeloceras* (about actual size), *below right*. Certain coiled ammonites of the Cretaceous had shells almost 7 feet in diameter – an all-time record for shelled invertebrates.

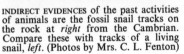

INDIRECT EVIDENCES of the past activities of animals are the fossil snail tracks on the rock at *right* from the Cambrian. Compare these with tracks of a living snail, *left*. (Photos by Mrs. C. L. Fenton)

AN ANCESTOR OF LIMULUS is *Prestwitchia* from the Pennsylvanian, a fossil member of the primitive arachnid-like group of which *Limulus* is the sole survivor. (About ⅔ actual size)

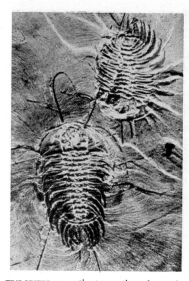

TRILOBITES are extinct aquatic arthropods allied to crustaceans. These two beautifully preserved fossils of *Neolenus* from mid-Cambrian rocks clearly show the trilobed body. Notice the antenna on the one at the left. (Photo courtesy American Museum of Natural History)

EURYPTERIDS are extinct aquatic arthropods related to arachnids. *Eurypterus,* from the Silurian, shows some small appendages and two large ones. The long segmented abdomen has no appendages and ends in a spine. (About ½ actual size)

RESTORATION OF A SILURIAN SEA BOTTOM shows a large eurypterid, *Pterygotus,* which reached a length of 9 feet, the greatest size known for any arthropod. These heavy animals probably spent most of their time on the bottom (whether in the sea or in fresh water is an open question), but the two large paddle-like appendages suggest that they could swim. They were probably carnivorous. Like the trilobites and many other invertebrate groups, they became extinct at the end of the Paleozoic. (Photo courtesy Buffalo Museum of Science)

THE LARGEST INSECT THAT EVER LIVED had a wing spread of 2½ feet. It belonged to the order Protodonata, extinct since the end of the Triassic period, which probably gave rise to modern dragon-flies. This restoration is of a specimen found in Permian rocks near Elmo, Kansas. (Photo courtesy Field Museum)

FOSSIL ANT from the Tertiary rocks of Colorado, U.S.A. Note the large compound eyes. (Photo from F. M. Carpenter)

Above. AN ANT imbedded in Tertiary amber (fossil resin) from the Baltic region. (Photo by E. Magdeburg)

TWO TERMITES, imbedded in amber from the middle Tertiary (about 38,000,000 years ago) look as if they had died yesterday. (Specimen lent by A. E. Emerson. Photo by P. S. Tice)

Above. TERMITE WING on a piece of Middle Tertiary rock from Spokane, Washington, U.S.A. Even the most delicate veins are easily seen. (Photo courtesy A. E. Emerson)

THE LOBSTER AND OTHER ARTHROPODS

successful on land – but only because they avoid certain problems by living in moist places. For truly successful land forms we must look to the other classes of arthropods.

ARACHNIDS

THE class ARACHNIDA includes the spiders, scorpions, mites and ticks, harvestmen ('daddy longlegs'), and a few minor groups. No other class of animals is less loved by most people. There is some basis for this dislike, in that scorpions and some spiders can inject a poison which produces painful, though usually not serious, results in man; some mites are parasites in human skin; and some ticks suck human blood and spread disease. But relatively few people in large cities have ever had a single unpleasant experience with an arachnid. The sinister reputation of a group like the spiders, which do little harm and some good (by killing insects undesirable to man), is based on nothing more than a vague fear of animals which have long legs, run rapidly, live in dark places, and catch their prey in webs or other traps.

Hardly any description will fit all the orders; but, in general, arachnids are terrestrial arthropods which have the body divided into two main regions: a CEPHALOTHORAX bearing six pairs of appendages, of which four of the pairs are walking legs, and an ABDOMEN which has no locomotory appendages, though it may have some other kind. The *four pairs of walking legs* usually serve as a convenient, if superficial, way of distinguishing arachnids from insects, which have only three pairs. But the difference between the groups is much more deep-seated. Arachnids differ from crustaceans and insects in having *no compound eyes*, only simple ones. And they are even more clearly marked off from crustaceans, centipedes, millipedes, and insects by the nature of the segmental appendages on the head. Arachnids have *no antennas*, the function of these organs being served by an abundance of sensory bristles or 'hairs' with which the body and particularly the appendages are covered. Also, they have *no true jaws* homologous with those of crustaceans and most other arthropods. None of the arachnid appendages are completely specialized for chewing, but on the basal segments of one or more of them are sharp biting processes. Many primitive arthropods have such chewing processes on the bases of the appendages, and it is thought that from such structures came the more specialized jaws of crustaceans or insects. In front of the mouth (on the third segment) arachnids have a pair of CHELICERAS, appendages which may take the form of pincers or of sharp, fanglike claws. Behind the mouth (on the fourth segment) is a pair of PEDIPALPS, leglike appendages that serve a sensory function, as in spiders, or are used for seizing prey, as in scorpions. Among the various arachnids either the cheliceras or the pedipalps are the important weapons of offence, but never both in the same animal.

The spiders are by far the largest and most widely distributed order of arachnids. A generalized description of a spider, though applying in many respects only to this one group, will give some further idea of arachnid structure and habit.

IN a SPIDER the cephalothorax is covered by a shield, the carapace, on which are set the simple eyes, usually eight in number. The cheliceras are sharp and pointed and are used for capturing and then paralyzing the prey by injecting a poison. Ducts from a pair of poison glands lead through the cheliceras and open near their perforated tips. The pedipalps look like legs but are sensory, and their basal joints have jawlike processes which hold and compress the prey. In the male the pedipalps are modified for transferring the sperms to the female. The four pairs of walking legs end in curved claws. The abdomen shows no external evidence of segmentation and has no appendages except the SPINNERETS, of which there are usually three pairs. The spinnerets are finger-like organs which have at their tips a battery of minute spinning tubes (sometimes a hundred or more on each spinneret), from which the fluid silk issues, and then hardens as it comes in contact with the air. The spinning tubes connect with several kinds of silk glands which produce different kinds of silk for spinning various parts of the web, making a protective cocoon for the eggs, binding the prey, etc. Some of the tubes produce not silk but a sticky fluid which makes the threads of the web adhesive.

When an insect or other small animal becomes entangled in the web, the spider apparently feels the tugging, for it hurries to the scene, seizes the struggling animal, and, holding it between the jawlike processes on the pedipalps, injects a poison. If the prey is large and formidable, the spider may use a more indirect method, first binding its victim with silk. The mouth of the spider is too small to swallow solid food; instead, the animal injects a digestive fluid through the wound made by the bite of the cheliceras. The predigested, liquefied tissues of the prey are then sucked up by means of a muscular SUCKING STOMACH (aided by the squeezing action of the pedipalps and the sucking action of the pharynx). Beyond the sucking stomach the digestive tract gives off several pairs of pouches, which increase the digestive and absorptive surface, and a large DIGESTIVE GLAND, which branches extensively and occupies most of the spider's abdomen. This gland is the main organ of digestion and is capable of taking up very large quantities of food at one time, storing it, and then gradually absorbing it. This enables spiders to go for long periods without taking food (though they must have water quite often).

The circulatory system is open, as in other arthropods. The heart lies dorsally in a large sinus and receives blood through openings in its sides. The excretory system, as in most other arthropods, consists of tubules which open into the intestine. In many spiders there are also excretory sacs which open near the bases of the legs. The respiratory organs are of two types. The LUNG BOOK is an air-filled sac which communicates with the external air through a slitlike opening. Attached to the walls of the sac is a series of leaflike folds of the body wall. These 'leaves', which have suggested the name of the organ, are held apart by supports so that air can circulate freely between them. The spaces within the leaves are filled with blood and communicate with the blood sinuses of the abdomen. Thus the leaves of the lung book are simply another device for exposing a large amount of respiratory surface to the air. The AIR TUBES of spiders receive air through openings on the abdomen and convey it to the tissues. The smallest tubes usually do not branch extensively, as in insects. The air tubes are not thought to be homologous in the two groups, for arachnids and insects probably did not have a common ancestor which lived on land. Some spiders have only lung books, and some have only air tubes, but most have one pair of lung books and one pair of openings to air tubes.

KING CRABS

THE king crabs are not crabs at all but primitive marine arthropods, the only living representatives of the class PALEOSTRACHA.

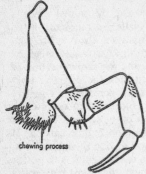

First walking leg of *Limulus*, showing the CHEWING PROCESS on its base.

There are five living species of king crabs, all of them usually placed in the single genus *Limulus*. These animals are often referred to as 'living fossils' because they have changed so little from the earliest fossil representatives of the group. No one can say with any certainty why they have been able, with no 'modern improvements', to survive in competition with more highly developed aquatic arthropods. Perhaps their success results from a combination of unobtrusive habits and a heavy hoodlike carapace which forms a complete roof over the body and all the appendages.

King crabs live in shallow water along sandy or muddy shores. They can swim through the water by flapping the appendages on the abdomen, but spend most of the time burrowing in the sand or mud for the worms and molluscs on which they feed.

Apart from their interest as archaic forms, the king crabs have attracted attention because they are clearly related to arachnids and, though somewhat specialized, give us some idea of what the ancestral aquatic arachnid may have been like. The king crabs are, in fact, often classified as one of the orders of the class Arachnida, their chief differences being associated with their aquatic life. Attached to the flattened abdominal appendages are the GILL BOOKS, groups of thin plates in which blood circulates; they are so similar in plan to the lung books of terrestrial arachnids as to suggest strongly that some sort of gill book was the forerunner of the lung book. As in arachnids, the body is divided into cephalothorax and abdomen; and the thorax has six pairs of appendages, of which the first is a pair of pinching cheliceras and the other five pairs are walking legs. The first four pairs of walking legs have on their bases spiny processes for masticating the food.

The eggs hatch into free-swimming larvas which lack the long tail of the adult.

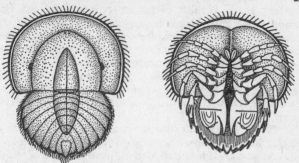

The free-swimming LARVA OF *Limulus*. *Left*, dorsal view; *right*, ventral view. (After Kingsley.)

CENTIPEDES

THE centipedes ('hundred-legged') form the class CHILOPODA. The members are land arthropods which are flattened dorso-ventrally and have a distinct head, followed by numerous similar body segments. The appendages of the first body segment (seventh segment of the animal) are modified as a pair of POISON CLAWS. These have perforated tips through which a poisonous secretion, from a pair of glands, can be injected into the prey. Each of the other body segments, except the last two, has a pair of walking legs. Some centipedes have as many as one hundred and seventy-three pairs of legs, and others have only fifteen, but thirty-five is probably an average number. The animals run very rapidly, and the numerous legs apparently work in perfect co-ordination.

The head, as in all arthropods, has six segments, the appendages of which are homologous with those of a lobster or insect but resemble more closely those of an insect. There is a single pair of antennas; the jaws have no sensory process or palp; the first maxillas have two lobes; and the second maxillas are usually fused together, as in insects. There are two groups of simple eyes on the head; but other parts of the body must be sensitive to light, for some centipedes react negatively to a bright light when the eyes are completely covered with heavy paint.

In other respects they are much like insects (described in chap. 24). The digestive tract is a straight, simple tube. The excretory system consists of tubules opening into the hind portion of the gut. Oxygen is supplied to the tissues through branched air tubes which lead from a pair of openings in every segment. The circulatory system is slightly more elaborate than in insects, having a pair of arteries to every segment.

Lithobius, a common centipede, lays its eggs in the ground. The young hatch with only seven segments, and the rest are added later. During growth the animal sheds its exoskeleton frequently.

Centipedes are found in moist situations under the bark of decaying logs and under stones. They are carnivorous, feeding upon soft insects such as cockroaches, plant lice, and silverfish; they also eat earthworms and slugs. They hunt only in the dark and are probably guided in their movements mostly by touch, to which they are very sensitive. They seldom come to rest unless the body is in contact with some solid object on at least two sides. This is adaptive, since it keeps these animals under cover, where they are safe from enemies and from drying.

Though terrestrial, such a centipede as *Lithobius* can survive many hours completely immersed in water, but will die in a few hours in an uncovered dish of dry earth. That their habit of keeping under cover is a positive reaction to contact as well as a negative response to light (which centipedes avoid when possible) can be shown by a simple experiment. A *Lithobius* placed in a glass dish will run about ceaselessly; but if some narrow, transparent glass tubing is placed in the dish, the animal will soon come to rest in the tubing, which affords a maximum of contact with the surface of the centipede.

MILLIPEDES

THE name millipede means 'thousand-legged'; and though this is a gross exaggeration, millipedes do have very large numbers of legs – in fact, twice as many as a centipede of about the same length, since there are two pairs to each adult abdominal segment. The technical name of the group is class DIPLOPODA, which refers to the double-legged situation. A millipede embryo has only one pair of legs to every body segment, each innervated by a segmental ganglion. In the adult the first four (thoracic) segments remain single, but the other (abdominal) segments fuse in pairs, so that each adult ring represents two embryonic segments and has two pairs of legs.

The head has six segments with the same appendages as in centipedes except that the first maxillas, which appear in the embryo, do not persist to the adult stage. The eyes superficially resemble compound eyes, but each is only a clump of many simple eyes set closely together. The internal anatomy resembles that of centipedes.

In contrast to the centipedes, which are all carnivorous, most millipedes are herbivorous, and in spite of their larger number of legs run much more slowly.

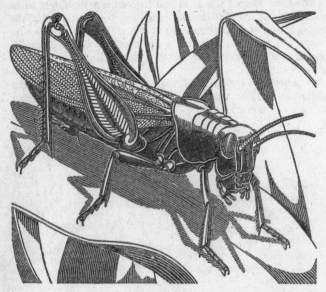

CHAPTER 24

The Grasshopper and other Insects

MOTORING across the hot, dry, desolate regions of the Western U.S.A., one is surprised not by the scarcity of human beings but by the fact that anyone at all should choose to live in the Mojave Desert, when life is much more pleasant in Los Angeles, only a short distance away. The answer is that men, like all animals, compete with the other members of their species for food and shelter. And since the most desirable regions are usually filled to capacity, those individuals who can adapt themselves to the conditions of the less favourable places have the advantage of a relatively unexploited environment.

The most successful animal groups are those in which the members do not all live in one region in competition with one another but have spread out to every corner of the world that will support their mode of life. The problems of why particular animals live where they do, how they reached there from the places where they first evolved, and why they do not occupy certain other regions which appear to be suitable for them, belong to a major subdivision of biology which we call *animal geography* and which we cannot adequately consider here. We can only mention, in passing, that arthropods, more than any other group of higher animals, are WIDELY DISTRIBUTED. They occur in all seas, in all bodies of fresh water, and in every land habitat. Of the many factors that make for wide distribution, one of the most important is small size; and arthropods are mostly small animals which can be carried from one place to another by water currents, by the wind, on floating debris, and on the bodies of other animals. Equally important is mobility; and of all arthropods, the most widely distributed are the winged insects, some of which are not stopped in their migrations even by barriers such as mountains or large bodies of water.

Ease of distribution is by no means the secret of insect success. For there is little advantage to animals in moving to new places if they cannot adapt themselves to the conditions of life there. Since the insect body plan does lend itself readily to structural specialization for almost any mode of life, insects have been able to spread from their original centres of distribution and are now the dominant invertebrates of tropical rain forest, temperate forest, prairie, plain, desert, and tundra. But even this is not enough. If all the insects in an oak wood were grass-eaters, they would be competing with one another for the limited amount of grass available. Instead, they show such a diversity of habits that they occupy every niche – and, in so doing, tap every source of energy available in that community. In the oak woods we would find insects living as predacious carnivores, herbivores, suckers of plant juices, suckers of vertebrate blood, pollen-gatherers, nectar-gatherers, scavengers, parasites of plants, and parasites of animals. Even within each of these major categories they do not all compete with one another. Some herbivores eat grass, some leaves of trees, some woody stems. Among leaf-eaters, some may specialize on particular species of plants. Among wood-eaters, some eat bark, some sapwood, and others only decayed wood. Of this last group, some are limited to wood in a particular stage of decay.

Thus we see that animals meet competition either by excelling their neighbours in one of the more typical modes of life, by becoming adapted to a relatively unexploited environment, or by becoming specialized to exploit some source of energy not available to the other members of a crowded community. This tendency of animals to spread out into every available niche in any habitat is called ADAPTIVE RADIATION. A few of the lower phyla, in which the simple body plan does not allow radical modifications, show practically no radiation and are limited to a single way of life. For example, all sponges are sessile forms

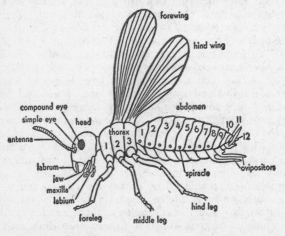

Diagram of a TYPICAL INSECT. Most insects have fewer abdominal segments, owing to loss or fusion at the posterior end. (After Snodgrass.)

which feed by drawing through the body a current of water from which they strain microscopic particles; all bryozoans and all brachiopods are attached forms which feed by ciliary currents. But this is not the rule. Most phyla, particularly the higher ones, which have complex body plans, show radiation not only within the phylum as a whole but also within a single class or even an order. In categories smaller than this there is little significant variation; for while the different genera of a family may live in widely separated parts of the world, where they do not come in competition with one another, usually they fill corresponding roles in their communities.

Some variation within the family does occur, as in the case of the bark-beetle family (Scolytidae). This group is composed mostly of wood-eaters which excavate tunnels just under the bark, but some of the members dig through the solid wood and then use the tunnels to grow fungi, on which they feed. This can hardly be classed as adaptive radiation, by which we imply the penetration of a group into practically all the major roles in any region.

Of all animals, arthropods best illustrate adaptive radiation. It is clearly shown among crustaceans, but is carried to the greatest extremes in the INSECTS. The name 'insect' comes from a Latin word meaning 'incised' and refers to the fact that insects generally have a sharp division between the head and thorax and between the thorax and abdomen. The head of the adult insect is all in one piece, but the six segments can be seen in the embryo, and some of them are indicated in the adult by the paired appendages. The thorax, which has undergone less specialization and fusion than the head, has three segments, each bearing a pair of appendages, usually walking legs. The second and third thoracic segments each bear a pair of wings (except in certain primitive and degenerate forms). The abdomen is the least specialized region of insects, being composed of relatively similar segments, generally without appendages, unless we consider certain structures at the posterior end of the abdomen to be modified segmental appendages. The external anatomy of insects varies so much that it is less satisfactory to generalize than to describe a particular insect such as a grasshopper, which is not too specialized and is therefore usually considered a fairly typical representative of its class.

THE GRASSHOPPER

THE HEAD of the grasshopper has two compound eyes and three simple ones (ocelli). The *compound eyes* are similar in structure to those of the lobster; and while they are not on stalks, they have a broad field of vision because they occupy a relatively large area and curve round the sides of the head. They occur on the first head segment, which has no segmental appendages. The second segment bears a pair of long, jointed sensory *antennas*, homologous to the first antennas of the lobster. The third segment bears the *upper lip* (labrum), which is not serially homologous to the other head appendages but is a secondary growth. The fourth has a pair of toothed horny *jaws* (mandibles). The

fifth is indicated by a pair of accessory jaws, or *maxillas*, each
with a jointed sensory palp. The sixth bears the *lower lip* (labium),
which has on each side a sensory palp. The labium of the adult
appears as a single plate, but in the embryo it arises as a pair of
structures which later fuse in the mid-line. Thus, each half of the
labium is homologous to one of the second maxillas of the
lobster.

The THORAX is partly covered dorsally by a chitinous shield
and by the wings; but the three segments, with a pair of legs

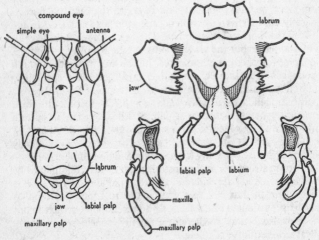

Left, front view of the HEAD OF A GRASSHOPPER. *Right*, the mouth parts removed from
their attachments. (After Snodgrass.)

on each, are clearly visible at the sides. The *legs* are composed of a
characteristic series of joints (as shown in the drawing) and end
in two curved claws between which is a fleshy pad that aids in
clinging to surfaces. The first two pairs of legs are typical walking
legs. The third is specialized for jumping; it has one joint, the
femur, which contains muscles for jumping, enlarged out of
proportion to the others. The *two pairs of wings* are different
from each other. The first is hardened and does not function in
flying; it serves only as a cover for the hind wings, which do the
actual flying. The hind wings are quite broad when in flight, but
when not in use are folded like a fan and fit under the first pair.

This arrangement is a specialization; in most generalized insects the two pairs of wings are more alike and both are used in flight. The wings are made of cuticle and are stiffened by thickenings called *veins*.

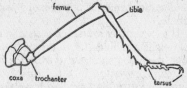

The MIDDLE LEG of the grasshopper shows, in relatively unspecialized form, the parts of a typical insect leg. (After Snodgrass.)

The ABDOMEN has no appendages except those at the posterior end, which are associated with mating and egg-laying. The abdomen contains much of the machinery of the body, since the head is small and the thorax is nearly filled with the muscles that move the legs and wings.

The DIGESTIVE TRACT consists of three parts: fore-gut, mid-gut, and hind-gut. The *fore-gut* starts at the mouth, receives a secretion from the salivary glands, and runs on as a narrow oesophagus, which leads to the *crop*, a large, thin-walled sac in the thorax. On the inner walls of the crop are transverse ridges armed with rows of spines which probably serve to cut the food into shreds. The crop is mainly a storage sac which enables the grasshopper to eat a large quantity at one time and afterwards digest it leisurely. From the crop the food passes into a muscular *gizzard*, lined with chitinous teeth. At the posterior end of the gizzard is a valve, which prevents the food from passing into the stomach before it is thoroughly ground and also prevents food in the stomach from being regurgitated. Digestion probably begins in the crop, for the food entering that organ is already mixed with salivary secretion, and it also receives some digestive juices which pass anteriorly from the stomach. Since the whole fore-gut is lined with cuticle, little, if any, absorption of food occurs there. The *mid-gut*, or stomach, which lies mainly in the abdomen, has no cuticular lining and serves as the main organ of digestion and absorption. Opening into the anterior end of the stomach are six pairs of pouches: one pouch of each pair extends anteriorly from the point of attachment, and the other posteriorly. These pouches secrete a digestive juice and also aid in absorption. The junction of the stomach with the *hind-gut*, or intestine, is marked by the attachment of long excretory tubules. The intestine is lined with cuticle. It receives the waste materials of digestion and the nitrogenous excretions of the excretory tubules.

The EXCRETORY SYSTEM consists of a number of tubules (called 'Malpighian tubules', from the name of their discoverer) which lie in the blood sinuses and from the blood extract nitrogenous wastes. The wastes, in the form of crystals of uric acid, are poured into the hind-gut and leave the grasshopper, by way of the anus, as dry excretions. Dry wastes are characteristic of small land animals, which have a limited supply of water.

Air for RESPIRATION is not distributed by the circulatory

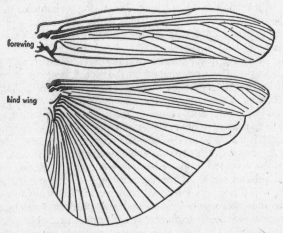

WINGS of the grasshopper. Only the larger veins are shown. (After Snodgrass.)

system but is piped through branching *air tubes* (tracheal tubes) which form a definite system of longitudinal and transverse main trunks from which smaller branches ramify to all parts of the body. The air tubes lead from paired openings which lie at the sides of the abdomen and thorax in the thin membrane between segments. The openings, or *spiracles*, of which there are ten pairs, are guarded by hairs to keep out dirt, and by a valve which can be opened or closed to regulate the flow of air. Their closure also aids in decreasing the evaporation of water. The air tubes are prevented from collapsing by means of spiral thickenings in their walls. They branch freely and become so small

(1 micron, or 1/25,000 of an inch, in diameter) that the finest ones (air capillaries, or tracheoles) are made of a single cell; the smallest tubes have no spiral thickening and are usually filled with fluid. Oxygen in the air of the larger tubes dissolves in this fluid and passes, by diffusion, to the tissues. Here and there the system widens into large air sacs. The air is moved chiefly by diffusion; but muscular breathing movements, which alternately compress the air sacs and then allow them to expand, aid in changing the air. The greater the muscular activity, the greater the pumping action on the air sacs and the better the circulation of air. In the grasshopper the first four spiracles open into one set of air tubes only at inspiration, and the remaining six pairs

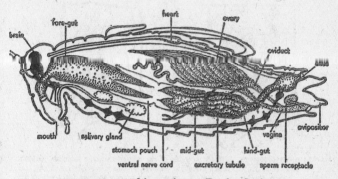

INTERNAL ANATOMY of the grasshopper. (Based on Snodgrass.)

are open only upon expiration; this facilitates the flow of air. The larger air tubes are impermeable to water (thus preventing water loss) but freely permeable to oxygen, which dissolves in the blood and has to travel only a very short distance to the tissues. There are no respiratory pigments for carrying oxygen. Carbon dioxide leaves by the reverse route but may also escape through the thin parts of the body surface.

The system of air tubes, as it occurs in the grasshopper and other insects, is one of the factors which limits the size of these animals. Since the air must travel mostly by diffusion, it could not reach the interior of a large animal fast enough to support the degree of activity displayed by insects. The other main factor in limiting the size is the chitinous exoskeleton, which in a larger animal would make flying more difficult.

The blood vessels of the CIRCULATORY SYSTEM are much less extensive than in the lobster. In fact, there is only one vessel, the long contractile *dorsal vessel*, composed of the tubular *heart*, which pumps the blood forward, and its anterior extension, the *aorta*. In each segment through which it passes, the heart is dilated into a chamber perforated on each side by a slitlike opening through which blood enters. The aorta carries the blood into the head and there ends abruptly. The blood flows out into spaces among the tissues and makes its way back into the thorax, where it bathes the thoracic muscles. From there it enters the abdomen and bathes the various organs, absorbing food from the stomach and giving up wastes to the excretory tubules. Then

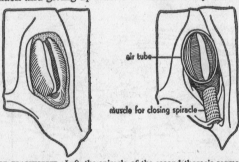

air tube

muscle for closing spiracle

SPIRACLE OF GRASSHOPPER. *Left*, the spiracle of the second thoracic segment, as seen from the outside. It consists of a vertical slit guarded by two valve-like lips. *Right*, the same viewed from the inside to show the muscle which pulls the two lips together to close the spiracle. (After Snodgrass.)

it returns to the heart. The course of the blood is really more definite than this brief sketch has intimated. There are partitions which deflect the blood so that it enters one side of each leg and emerges on the other. In the abdomen there are two large horizontal partitions which aid in directing its course, and the dorsal one separates the cavity containing the heart from that in which the other viscera lie. In an open system of this kind, where the blood flows among the tissues instead of in definite vessels, the rate of flow is relatively low. However, this is no disadvantage, since the distribution of oxygen has been taken over largely by the system of air tubes. The blood serves mainly as the distributing and collecting medium for food and wastes, but it has other functions. It acts as a reservoir for food; and, when under pressure, it aids in hatching from the egg, in moulting, and in the expansion of the wings. Besides, it contains cells which ingest bacteria and wall off parasites.

The NERVOUS SYSTEM is a ventral, double, ganglionated cord like that of the lobster. The embryo has a ganglion for each segment, including six in the head; but in the adult some of the ganglia are fused so that there are only two in the head, three in the thorax, and five in the abdomen. Still, this is a fairly generalized system, as compared with that of some insects, which have all the thoracic and abdominal ganglia fused into one mass. The *brain* lies above the oesophagus and between the eyes. It is joined to the first ventral ganglion by a pair of nerves which encircle the gut. The brain has no centres for co-ordinating muscular activity; after removal of the brain the animal can walk, jump, or fly. As in lower invertebrates, the brain serves chiefly as a sensory relay

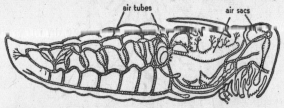

RESPIRATORY SYSTEM of a grasshopper, showing some of the main air tubes and air sacs. (After Vinal.)

which receives stimuli from the sense organs on the head and, in response to these stimuli, directs the movements of the body. It also exerts an inhibiting influence, for a grasshopper without a brain responds to the slightest stimulus by jumping or flying – a very unadaptive kind of behaviour. And even in the absence of any external stimulation, the animal displays an incessant activity of the palps and legs. The *first ventral ganglion* controls the movements of the mouth parts and plays some role in maintaining balance. The *segmental ganglia* are connected and co-ordinated by nerves which run in the cords, but each is an almost completely independent centre in control of the movements of its respective segment (or segments) and appendages. In some insects these movements have been shown to continue in segments which have been severed from the rest of the body. An isolated thorax is capable of walking by itself, and an isolated abdominal segment performs breathing movements.

The REPRODUCTIVE SYSTEM of the male grasshopper is a pair of

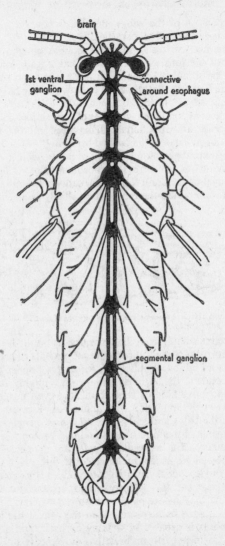

NERVOUS SYSTEM of the grasshopper. (Modified after Snodgrass.)

testes, which discharge sperms into a *sperm duct*. The duct receives secretions (a fluid in which sperms are conveyed to the female) from glands and then opens near the posterior end of the body. In the female

GRASSHOPPER LAYING EGGS in the ground. On the right is a completed batch of eggs. (After Walton.

there is a conspicuous set of stout appendages near the posterior end of the abdomen. These are used for digging a hole in the ground in which to lay the eggs. Internally there is a pair of *ovaries*, with *oviducts*. The two oviducts converge into a *vagina*, which opens into a *genital chamber.* In the sex act the male introduces sperms into a special sac in the female, the *sperm receptacle*, in which they are stored until the time for egg-laying. The mature eggs pass down the oviduct and, contrary to the usual procedure in animals, the shell and yolk are put on before fertilization. A small pore is left, however, through which the sperm enters. As the eggs pass into the genital chamber, they are fertilized by sperms ejected from the sperm receptacle.

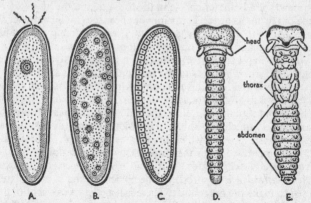

DEVELOPMENT OF AN INSECT. A, fertilization. B, the nuclei migrate to the periphery. C, cell walls appear between the nuclei, resulting in a single-layered embryo which corresponds to the blastula of other animals. D, when segmentation first appears, the head is composed of only three segments. E, later, the first three body segments are added to the head, forming a six-segmented head, which in the adult shows little trace of the original segmentation. (After Snodgrass.)

GRADUAL METAMORPHOSIS of a grasshopper from first nymphal stage to adult. Between any two successive stages the animal moults. The first nymph has a relatively large head and no wings.

The DEVELOPMENT of insects is rather different from that of most animals – chiefly because of the large amount of yolk, which provides enough food to enable the young to develop sufficiently to be able to feed. Instead of dividing into two cells and then into four, and so on to form a blastula, the zygote nucleus divides many times without the division of the cytoplasm. The nuclei then move to the periphery, and cell walls appear between them, thus forming a layer of cells. Subsequently, cells pass inward until there are roughly three layers. The outermost layer is the ectoderm; the endoderm and mesoderm are interior to it. Segments eventually appear as little hollow blocks of mesoderm. The mid-gut develops from endoderm. The foregut and hind-gut develop from infoldings of the ectoderm. Since the ectoderm alone secretes cuticle, this explains the presence of a lining of cuticle in the fore-gut and the hind-gut. A coelom forms as a series of pairs of hollow sacs in the mesoderm, one pair for each body segment. These coelomic sacs later break down and do not form the adult body cavity.

The young grasshopper, known as a NYMPH, hatches from the egg in a form which resembles the adult except in the relatively large size of the head, as compared with other parts, and in the lack of wings and reproductive organs. It feeds upon vegetation and grows rapidly; but since the chitinous exoskeleton cannot stretch very much, the animal must moult at intervals. MOULTING is a complex process which involves several steps. The outside layer is separated from that beneath by the secretion of a fluid from certain skin glands. This fluid dissolves part of the exoskeleton, which is absorbed by the epidermis and is presumably used to make the new cuticle. There is usually a weak spot in the skeleton down the middle of the back which breaks and provides an opening for the nymph to escape. The cuticle is finally ruptured, not by the increased size of the animal due to growth, which merely stretches the membranes between the segments, but by muscular contractions which build up great pressures on

In the adult the head is smaller in proportion to the rest of the body, and the wings are fully developed. (Combined from several sources.)

the skin. To aid in increasing temporarily the size of the body, the insect swallows air and closes the spiracles. The newly-emerged insect is soft and white; and since it is in a precarious condition, it usually retires to a safe place until the soft cuticle hardens and darkens. With each successive moult, of which there are five in most grasshoppers, differentiation continues, so that the final moult results in the adult grasshopper.

VARIATIONS IN INSECT STRUCTURE

To have observed insects at all is to have noticed that they vary tremendously – from flattened, crawling cockroaches, which feed on scraps of food, to flying butterflies, which suck nectar from flowers, and swimming beetles, which chase animal prey. These differences are mostly in external structures; internally, insects are more alike.

Variations in the DIGESTIVE TRACT are related mostly to what the animals eat. In the cockroach, which feeds on solid food, the gizzard is well developed, and its lining is armed with hard plates and spines. Insects which suck juices have no gizzard. In the honeybee the nectar is sucked up into a honey stomach, which corresponds to the crop of the grasshopper. The region between the honey stomach and the stomach (corresponding to the gizzard of the grasshopper) is a valve which prevents the food in the honey stomach, designed for storage in the hive, from going into the stomach.

Most insects obtain air for RESPIRATION through a system of air tubes; but some, like most of the tiny collembolans, have no air tubes and breathe through the body surface. Many insect larvas, which live as parasites in the fluids and tissues of their hosts, are equipped with well-developed systems of air tubes but must obtain oxygen by diffusion through the thin body wall. In many aquatic nymphs and larvas the spiracles do not function, and oxygen diffuses into the body through gill-like expansions

'Garden fleas' are primitive WINGLESS INSECTS (order Collembola) which never have had any winged ancestors.

of the body wall which contain air tubes (tracheal gills). Some aquatic larvas have thin-walled expansions of the body wall or extensions of the hind-gut which do not contain air tubes. In the absence of certain information we assume that oxygen diffuses into blood contained in these structures (blood gills) and so finally reaches the tissues.

There are also differences in the ganglia of the NERVOUS SYSTEM. In practically all insects there are two nervous masses in the head, but the number in the thorax and abdomen is more variable and depends upon the degree of fusion of the ganglia in these regions. The most extreme case of fusion is that of certain fly larvas, which have the entire ventral cord, including the first ventral ganglion, consolidated into a single mass.

The EXCRETORY TUBULES vary in number from two to over a hundred, but they all function in much the same way.

The essential parts of the REPRODUCTIVE ORGANS of insects are as described for the grasshopper. Hermaphroditism is known to occur in one species (a cottony-cushion scale insect) in which the females are able to fertilize their own eggs. Unfertilized eggs give rise to males, which are rare. The males have been seen to mate with the females, but whether they can fertilize the eggs is not known. In some species no males have ever been found, and the eggs laid by the female develop parthenogenetically (without fertilization). In the vast majority of insects, sperms are stored in a receptacle of the female at the time of mating, and the eggs are usually fertilized, as they issue from the oviduct, at the time of laying.

No matter how conservative they may be in their internal anatomy, the insects, as a group, show the most radical modifications and the greatest amount of external variation known for any class of animals. There are, of course, the easily observed differences in shape of body, colour, and size. But among the variations which are most important in adapting the animals to their different ways of life are those of the sense organs and

appendages. From an ancestral arthropod with numerous segments each bearing a pair of appendages which were all alike
and primarily locomotory in function has been derived a vast
array of insects which show a reduction and fixation of the
number of segments, a loss of appendages on the abdomen, and a
specialization of the appendages of the anterior part of the body
into a series of structures which are, at least on the head, all
different from one another. Moreover, in the different insect
groups corresponding appendages have been modified in a great
number of ways to fit the various animals to their particular
niches.

Insects generally agree in having as sense organs a pair of compound EYES, three simple eyes, and a pair of antennas. In addition,
the mouth parts may bear jointed sensory projections, the palps,
and the body is clothed with a variety of sensory hairs, scales,
pits, etc. There may also be special organs of smell or hearing.

The simple eyes probably do not form images, but act to increase the
sensitivity of the brain to light stimuli from the compound eyes. For,
if the three simple eyes are painted over with an opaque substance,
the insect does not react to light as rapidly as if the simple eyes were
not covered. If the large eyes are covered, the insect does not respond
to light. In insects which learn readily, colour vision can be demonstrated. In one type of experiment a table is put near a beehive, and on
the table are placed cards of different colours. On each card is set a
glass vessel filled with water, and sugar is added to the water in one
vessel – say the one on the blue card. In its excursions a bee finds the
sugar water and, while busily feeding, is marked with paint, so that it
can be recognized. After the bee has made several trips between the
table and the hive, the sugar water is switched to the yellow card. The
bee then returns to the blue card as before, even if the card is moved to
another position on the table, showing that the bee is reacting to colour
and not to position or odour. Similarly, a bee can be trained to respond
to yellow or to ultra-violet. Bees trained to red or black cannot dis-
criminate between these two colours, or between them and dark grey.
However, some insects do respond to red. If bees are trained to visit
blue, and then blue is replaced by grey on which is set a yellow card,
the bees now respond to the grey as if it were blue, apparently because,
as in the case of man, blue and yellow are complementary colours and
grey, set next to yellow, appears blue.

The ANTENNAS may be very long, as in crickets, cockroaches,
and katydids. They are the chief organs of TOUCH, for when the
antennas are removed cockroaches can no longer be trained to

turn right or left (see pp. 322-3). The touch receptors of the antenna are the fine hairs with which it is clothed. The hairs are stiff and are joined by a very delicate cuticle at their bases to the rest of the antenna. The antennas also bear organs for the sense of smell.

Experiments which demonstrate the sense of smell are similar to those on colour vision. First, it is necessary to determine if the insects react to odours. Sugar water is placed in small boxes, and, after bees have found them and are making trips to and from the hive, the box is replaced by one just like it, also containing sugar water, but sprinkled inside with flower extract. After the bees have made sufficient trips to get used to the scent, several new unscented boxes are placed beside a new scented one. When the bees return for more sugar, they buzz about the openings of the boxes but finally go inside the scented one. Further, when they are trained to go to one odour – say rose – they will not go to another, such as lavender. That the sense organs are on the antennas is shown by removing parts or all of the antennas from bees trained to certain scented boxes. When the last eight segments are removed from each antenna, the bees cannot distinguish odours. That this result is not due to the shock of the operation is proved by a control experiment in which some bees are first trained to visit blue boxes for sugar water. Then their antennas are removed, and it is found that they still return to the correct boxes. It may be noted, however, that with their antennas removed the bees have difficulty in entering the boxes, probably because of the partial loss of their tactile sense.

In some insects, as in dragonflies, to which sight is more important, the eyes occupy nearly the whole head, while the antennas are relatively minute. Scattered over the bodies of most insects are numerous TACTILE HAIRS. Other sense organs may occupy rather unusual places. For example, the SOUND-PER-CEIVING ORGANS of the grasshopper are on the sides of the abdomen just above the base of the third legs, while those of the katydid are near the upper end of the tibia of the first pair of legs. TASTE ORGANS, which enable the insect to distinguish sweet, salty, sour, and bitter, occur not only in the mouth but also on the antennas, palps, and feet. But just which of the several kinds of sense organs are the taste organs is not known.

The grasshopper, representing a fairly generalized group of insects, has *biting* MOUTH PARTS. These are the most primitive kind, and they are present also in beetles and in many other orders of insects. Two other main types of mouth parts are common: *sucking* mouth parts, as in butterflies; and *piercing and sucking*, as in the cicadas. In most butterflies the jaws are rudimentary, and the two maxillas are greatly elongated, each forming a half-tube, so that when they are held together they form the long sucking proboscis of the adult, through which liquids are pumped up by the mouth pump in the forepart of the head. The proboscis is extended only when the insect is feeding; when not in use, it is coiled under the head. The piercing beak of the cicada consists of the mandibles, maxillas, and labium, all thought to be homologous to the mouth parts of the grasshopper.

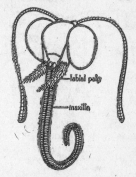

Head of a butterfly, showing SUCKING MOUTH PARTS.

The labium is a long tube but is not inserted into the food. It serves only as a sheath for the other mouth parts, being grooved on its dorsal surface to form a channel in which lie the mandibles and maxillas, which do the piercing. The mandibles are long, fine stylets with minute teeth at the end. The maxillas are similar but hooked at the tips; each is crescent-shaped in cross-section, and the two are fastened together by interlocking grooves and ridges to form a channel through which the food is drawn up by the sucking action of a muscular pharynx. Biting flies, mosquitoes, fleas, and bed-bugs have the mouth parts adapted in various ways for sucking blood. Other insects have still other modifications of the mouth parts, such as the sponging tongue of the housefly.

The thoracic LEGS of insects are modified in a variety of ways, but all are composed of the same basic parts (figured and named in the diagram of the grasshopper leg). Land forms have walking legs, with terminal pads and claws, as in the grasshopper, for clinging to vegetation or other objects. Houseflies have sticky pads at the tips which enable them to walk up smooth vertical surfaces, such as glass. Water beetles have flattened legs, fringed with bristles, for swimming. But the legs may serve other functions besides locomotion. The walking legs of the HONEYBEE are modified for collecting food. Each is highly specialized and quite different from the others, so that, together, they constitute a complete set of tools for collecting and manipulating the pollen upon which the bee feeds.

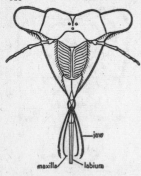

Head of a cicada, showing PIERC-
ING AND SUCKING MOUTH PARTS.

The FIRST LEG has many branched feathery hairs for collecting pollen. Along one edge of the inner surface of the tibia is a fringe of short, stiff hairs which form an *eye brush* used to clean the compound eyes. The large first joint of the tarsus is covered with long, unbranched hairs, forming a *pollen brush* for collecting the pollen grains that become caught among the hairs of the fore part of the body when the bee visits flowers. This first joint also has a semicircular notch lined with a comblike row of bristles and known as the *antenna comb*. The antenna is cleaned of pollen by drawing it through the notch. As it is pulled through, it is held in place by a spur on the end of the tibia which fits against the tarsal notch. Comb and spur together are called the *antenna-cleaner*. The MIDDLE LEG is the least specialized of the three. The large first tarsal joint is wide and flat and covered with stiff hairs which form a brush for removing pollen from the first legs and the thorax. On the lower end of the tibia is a spine, the *spur*, for removing the plates of wax from the wax glands on the ventral side of the abdomen. The HIND LEG is the most specialized, being fitted to carry the load of pollen. Rows of *pollen combs* on the inner surface of the very large and flattened first tarsal joint scrape the pollen from the second legs and posterior part of the abdomen. A series of stout spines, the *pecten*, on the lower end of the tibia removes the pollen from the combs of the opposite leg, and it falls on the *auricle*, a flattened plate on the upper end of the first tarsal joint. The leg is then flexed slightly, so that the auricle is pressed against the end surface of the tibia, compressing the pollen and pushing it on to the outer surface of the tibia and into the *pollen basket*. The pollen basket is formed by a concavity in the tibia, which has, along both edges, long hairs that curve outwards. Pollen clings together and to the basket hairs because it is moist with secretions from the mouth. When the baskets are loaded, the bee returns to the hive and deposits the pollen in special wax cells. Properly combined with sugar and other substances, the pollen mixture become 'beebread'. This provides a source of protein for both adults and larvas.

The WINGS of insects are flattened, two-layered expansions of the body wall and at first consist of the same parts: cuticle and epidermis. Later the two opposing layers meet and the inner ends of their cells unite, except along the channels in which lie nerves, air tubes, and blood spaces. In later stages the epidermal cells condense along these channels, forming the 'veins' of the wing. The epidermal cells finally degenerate, and the adult wing is almost completely made of cuticle, though it may have a circulation of blood and some sense organs on the surface.

Not all insects have wings. The two most primitive orders (Thysanura and Collembola) have never developed them. In some of the more specialized orders one or both pairs have been lost secondarily; the flies have only the first pair, the second being reduced to a pair of knoblike structures; and the fleas and

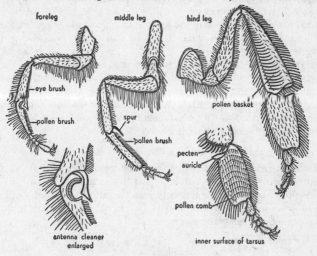

LEGS OF THE HONEYBEE.

lice have lost both pairs. Besides, among the orders which typically have two pairs of wings, there are wingless members. In the social insects certain castes lack wings; and in some species the males have wings while the females are wingless, or the opposite.

Typical wings are membranous structures stiffened by veins, and in many groups both wings are of this type and function in flight. In the grasshopper we saw that the first pair of wings were leathery, not used in flight, and served only as covers for the hind wings. In beetles the front wings are still more specialized, being only stiff horny plates which serve to cover the hind wings and much of the dorsal part of the body. In the less specialized orders the wings have a dense network of veins, as in dragonflies. In the higher orders, as in bees, there are only a few large veins. Since some wings in every insect group can be reduced to a common basic pattern of wing venation, it is thought that they are all homologous, having been derived from a common winged ancestor.

The ABDOMEN of insects bears appendages only at the posterior end; these are modified for mating, or as ovipositors, sensory projections, or in other ways. In the female grasshopper they are used for digging a hole in the ground in which to lay eggs. In the ichneumon flies the ovipositor is long and sharp; and when, in some way, the ichneumon fly senses the presence of a beetle larva within a tree, the ovipositor is used to drill a hole in the wood and deposit eggs in the body of the larva. When the eggs hatch, the young ichneumon larvas parasitize the beetle larva. In the honeybee, appendages at the posterior end are modified as a sting connected with poison glands, as some of us know from painful encounters with these insects.

METAMORPHOSIS

IN most insects the eggs develop after leaving the female, but some forms retain the developing eggs and give birth to fully-developed larvas. In the tsetse fly even the larva is retained within the body of the mother and is fed from special glands; on emerging, it is ready to pupate.

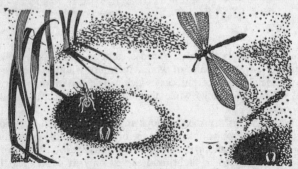

The ANT-LION is a flying insect with biting mouth parts. Its name is derived from the larva, which digs a pit in the sand and lies buried, with only the jaws protruding and held ready to grasp the first ant or other small arthropod that falls into the pit.

Not all insects undergo a development which involves changes radical enough to be termed a 'metamorphosis'. The thysanurans and the collembolans ('garden fleas', shown in the drawing on page 314 hatch from the egg in practically the same condition as the adult. They grow larger and later add joints to the antennas and some other appendages, but such changes are no greater than those undergone by most animals in their development. Some of the more specialized orders, like the lice, have secondarily lost the metamorphosis. The grasshopper is an example of an insect with incomplete metamorphosis. But since the nymph lives in the same habitat as the adult and eats the same food, this type of metamorphosis is often called GRADUAL METAMORPHOSIS, to distinguish it from the more thoroughgoing changes of the INCOMPLETE METAMORPHOSIS undergone by dragonflies and mayflies. These have immature stages which resemble the adult in general body form but have adaptive modifications which fit them to live in another habitat and eat food different from that of the adult. The dragonfly adult is terrestrial and catches other flying insects. The young form, or NAIAD, lives in the water and feeds on aquatic animals. The most striking changes of all are those undergone by the insects with COMPLETE METAMORPHOSIS, in which the feeding larval stage is very different from the adult not only in habit but also in details of structure. The most familiar example of this is the caterpillar and the butterfly. But not all larvas are creeping, wormlike forms which eat vegetation. The ant-lion larva has a broad, flattened body and digs a pit in the sand. It lies buried just below the centre of the pit, with only the pair of large pincer-like

jaws protruding (as shown in the drawing which heads this section). When an ant or other wingless insect stumbles into the pit, the sand slides under its feet, carrying it down into the waiting jaws of the larva. The adult is a winged form related to the lacewings. In complete metamorphosis the feeding larval stage and the adult are separated by a quiescent pupa stage in which the body of the larva is almost completely broken down and then reorganized. Since there is some tissue reorganization during the moulting of the grasshopper and other

The LARVA OF THE ANT-LION has long curved jaws and maxillas which form a pair of pincer-like structures for grasping prey and sucking its juices.

insects with a gradual or incomplete metamorphosis, the changes in the pupa may be interpreted as a more extreme type of moulting. There is some evidence that both metamorphosis and moulting are controlled by hormones, chemical substances secreted in one part of the body and then carried in the blood to other regions, where they initiate changes.

The bug *Rhodnius* feeds upon blood and requires only a single meal between each growth period after which it moults. If a bug is decapitated a few days before it is expected to moult for the last time (before becoming adult), it will still moult; but if it is decapitated early in the last growth period, it fails to moult, though it will live for months. This indicates that the head influences the moulting. Now, if the blood of an insect decapitated just before it is expected to moult is transfused into an insect decapitated very early in the growth period, the latter frequently moults. Apparently, then, there is a hormone, produced by glands somewhere in the head, which controls the moulting process.

The development, in the larva or other young stages, of structures which do not occur in the adult or in the ancestors of the species is called CENOGENESIS or 'sidewise evolution'. The pit-digging instincts of the ant-lion larva and the modification of its jaws as large pincers for seizing prey are specializations of the young stage which adapt it to its peculiar way of life. It is likely that no adult ant-lion ever dug pits in the sand or had such large pincer-like jaws. The specializations of the mosquito larva or the dragonfly nymph which enable them to live in the water, where the adult would only drown, are examples of the same thing. Of course, all structures which adapt the larva to its environment but are absent in the adult cannot be said to be cenogenetic. Many of them have been inherited from the racial history of the group. For example, the chewing jaws of a caterpillar enable it to feed on vegetation, whereas the adult has a long proboscis, representing a modification of the maxillas, for sucking nectar. The jaws can hardly be considered as 'sidewise evolution' of the young stage, for they arise as the appendages of the fourth head segment, and we know that primitive insects are all characterized by such simple chewing jaws.

BEHAVIOUR IN INSECTS

THE behaviour of insects is mostly instinctive and stereotyped, but a limited amount of learning is possible. Grasshoppers have been taught very little, but cockroaches can be trained to go to the right or left by a simple method of punishment for making the wrong turn. As was mentioned before, in connection with colour

SOCIAL INSECTS are equipped, from the start, with a set of instincts which fit them to their roles in the life of the colony. Here we see termite soldiers (with large jaws) guarding the colony while the workers carry on the other activities.

vision and smell, bees can be taught many things. In most species there is, of course, a difference in the instinctive behaviour patterns of the two sexes, and the female is equipped with a set of complex habits connected with deposition and protection of the eggs. In the social insects there is even more differentiation, each caste inheriting a set of instincts which fit it to its role in the life of the colony.

THE ORDERS OF INSECTS

THE class INSECTA is divided into more than twenty orders, the exact number depending upon whether certain closely similar

The DUNG BEETLE is a scavenger. It collects excreta of other animals. Here the beetle is shown using its hind legs to push a ball of dung toward its burrow.

groups should be placed in separate orders or lumped together in one. Within each order various families are modified to follow almost every insect 'walk of life'. Among beetles, for example, there are leaf-eaters, grain-feeders, wood-eaters, predators, scavengers (like the dung beetle shown in the drawing which heads this section), parasites, and commensals, to mention a few. In spite of all this diversity, the members of an order have in common the same general structure, including similar wings, mouth parts, and other appendages; and they also show the same type of metamorphosis. In spite of differences in shape, size, and colour, they have enough features in common for typical members of an order to be classified at a glance as flies, dragon-flies, termites, and so on. In the photographs in this chapter, the important orders of insects are presented, with illustrations of some of their features, their common representatives, and a few of the ways in which they affect the life of other animals, including man.

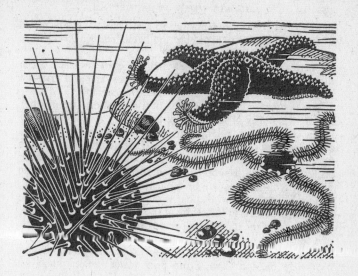

CHAPTER 25

Spiny-skinned Animals

WITH the arthropods we reached the peak of invertebrate specialization and should have closed our story. Unfortunately, organic evolution has followed none of the rules of good dramatic style and is full of anticlimaxes. There is one more major phylum to be described, the ECHINODERMATA ('spiny-skinned'), a group which very early diverged from the other main lines of evolution. The echinoderm body plan is utterly different from that of any other phylum; and the best clue to its relationship to other groups is the larva, which resembles the larva of certain primitive members of the phylum to which man and the other vertebrates belong. Of this, more will be said in the next chapter.

The phylum is exclusively marine and is divided into five classes: starfishes, serpent stars, sea urchins, sea cucumbers, and sea lilies. The echinoderm most familiar to everyone, even to those who have never been near the seashore, is the starfish.

A STARFISH has no resemblance to a fish and should be called a 'sea star', from its shape; but the name 'starfish' has gained wide usage and is hard to change.

There are a little over one thousand species of starfishes, which differ in certain details but as a class are more similar in structure and habit than the crustaceans or insects, for example.

The body of a starfish consists of a central disc from which radiate a number of ARMS. There are usually *five* of these, but many species have a larger number. Because of their shape, starfishes were once classed with the coelenterates; but now we know that their radial symmetry has nothing to do with that of coelenterates but has been secondarily derived from a basic bilateral symmetry, as will be explained later. There is no head, and the animal can move about with any one of the arms in the lead. The MOUTH is in the centre of the disc on the under surface.

The body of the starfish appears to be quite rigid but is capable of a considerable amount of bending and twisting. The rigidity is provided by a meshwork of CALCAREOUS PLATES which are embedded in the soft flesh. From these plates project numerous CALCAREOUS SPINES, some of them movable. The flexibility results from the fact that the plates are not united into a single shell but are distinct from one another and joined by connective tissue and muscles. The skeleton is like our own in that it is an *endoskeleton* embedded in the flesh, and differs from the exoskeleton of arthropods, which lies outside the body. It does not permit the freedom of movement of vertebrate or arthropod skeletons; but this is not important, as the starfish cannot move rapidly anyway.

LOCOMOTION in the starfish is by means of a kind of hydraulic-pressure mechanism, known as the WATER VASCULAR SYSTEM, unique to echinoderms. Water enters this system through minute openings in the SIEVE PLATE (madreporite) on the upper surface and is drawn, by ciliary action, down a tube, the STONE CANAL (so-named because its wall is stiffened by calcareous rings), to the RING CANAL, encircling the disc. From the ring canal arise five RADIAL CANALS, one for each arm. Each of these connects, by short side branches, with many pairs of TUBE FEET, hollow, thin-walled cylinders which end in suckers. Each tube foot connects with a rounded muscular sac, the AMPULLA. When the ampulla contracts, the water, prevented by a valve from flowing back into the radial canal, is forced under pressure into the tube foot. This extends the elastic tube foot, which attaches to the substratum by its sucker. Next, the longitudinal muscles of the tube foot contract, shortening it, forcing the water back into the ampulla and drawing the animal forward. Of course, one tube foot is a very weak structure; but there are hundreds of them, and their combined effort is capable of moving the starfish slowly over the ocean floor. The tube feet can work as described only when the animal is moving over a rock or some kind of hard substratum. On soft sand or mud the suckers are of little use and the tube feet act as little legs. Moreover, some starfishes have no suckers at the

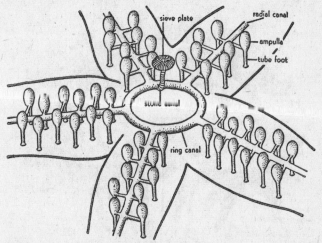

WATER-VASCULAR SYSTEM of starfish, showing essential structures. The ring canal may have one or more sets of vesicles; but as their exact function is not well known, they are omitted here.

ends of the tube feet. In the diagram of the cross-section of the starfish arm it can be seen that the ampullas lie within the body but that the radial canal and tube feet lie outside the calcareous plates in the centre of the under surface of the arm in a V-shaped groove (usually called the 'ambulacral groove', from a Latin word meaning a 'walk', because the double row of tube feet reminded someone of a flower-lined garden walk). On either side of this groove are rows of movable spines, which can be brought together to close the groove and protect the tube feet when the animal is attacked.

Starfishes move rather slowly; but they have no difficulty in running down their prey, for they FEED mostly on clams, which move still more slowly, or on oysters, which do not move at all. Anyone who has ever tried, barehanded, to open a live clam or oyster will wonder how it can be done by a starfish which is not much bigger than the clam it attacks. The starfish mounts upon the clam in a humped-up position, attaches its tube feet to the two shells, and begins to pull. The clam reacts to this by closing its valves tightly. The starfish uses the numerous tube feet in relays,

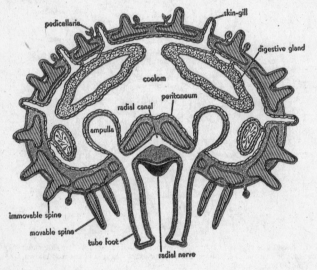

CROSS-SECTION OF ARM OF STARFISH to show large coelom, completely lined by a meso-dermal epithelium.

and is therefore able to outlast the clam, maintaining its pull on the shells until the two large muscles which hold the valves to-gether become fatigued and finally relax. When the shells gape, the starfish turns the lower part of its STOMACH inside out and extends it through the mouth, enveloping the soft parts of the clam and digesting them. The partly digested material is taken in-to the stomach and then into the five pairs of digestive glands (one pair in each arm) which connect with the upper part of the stomach. Very little indigestible material is taken in by this

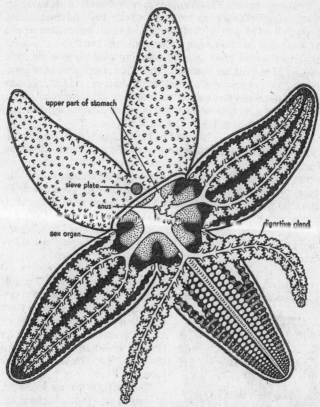

STARFISH, with body wall dissected away from the dorsal surface of three arms and most of the disc, to show digestive system. In the arm on the lower right, the two branches of the digestive gland have been spread apart to show the rows of ampullas and the sex organs. The upper part of the stomach is unshaded; the lower part, stippled.

method of feeding, and this accounts for the fact that there is practically no intestine and that the anus, a very small opening near the centre of the upper surface, is used very little, if at all. When the starfish eats a small clam, it may take the whole animal into the stomach; the shell is later ejected from the mouth.

Running beneath the water vascular system are fluid-filled channels of uncertain function, but there is no well-organized circulatory system. The distribution of materials is provided for by the large COELOM lying between the body wall and the digestive tract and filled with a fluid which bathes all the organs and is kept moving by cilia on the coelomic lining. The thin walls of the tube feet permit gaseous exchanges, but they are not the main organs of respiration. Opening from the coelom are many small finger-like projections, the SKIN-GILLS, which extend from the surface of the body through spaces between the calcareous plates. They have very thin walls through which oxygen diffuses into the coelomic fluid and carbon dioxide diffuses out.

The delicate skin-gills would be in a dangerously exposed position if it were not that they lie among the heavy spines and are further protected by the PEDICELLARIAS, small pincers which occur in the spaces between the spines or in clumps around the bases of spines. When a small animal creeps over the surface of a starfish, it is caught and held by the toothed jaws of the pedicellarias, which, together with currents created by cilia of the surface epithelium, serve to protect the skin-gills and keep the surface free of any debris which would interfere with respiration and excretion.

There is no specialized excretory system. EXCRETION is accomplished by amoeboid cells in the coelomic fluid which engulf nitrogenous wastes and then escape through the walls of the skin-gills. Probably also excretory are two branched pouches (shown in the diagram of the digestive system) which lie in the coelom and open into the intestine near the anus.

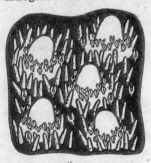

Small portion of the SURFACE OF A STARFISH to show the delicate *skin-gills* occupying the spaces between the large protective *spines*. The tiny *pedicellarias* occur in 'rosettes' around the bases of the spines and also among the skin-gills.

The NERVOUS SYSTEM is simple, as in most slowly-moving animals. A ring of nervous tissue encircles the mouth and gives off five radial nerves which extend along the middle of each arm below the radial canals of the water vascular system. There is also a deeper-lying ventral system and one near the upper surface. The SENSE ORGANS are poorly developed. Sensory cells, which occur all over the surface, are sensitive to touch. At the tip of each arm is a pigmented eyespot and a short tentacle, thought to be sensitive to food or other chemical stimuli.

TWO OVARIES OR TESTES lie in each arm and open directly to the exterior through small pores in the angles of the arms. The eggs and sperms of starfish, like many other marine animals, are shed into the sea water, where fertilization occurs.

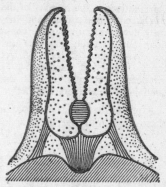

PEDICELLARIA. (After Cuenot.)

Since the eggs and sperms are more easily obtained from mature starfish and sea urchins than from most other animals, they have been much used by embryologists to study the mechanisms of fertilization and early development. Eggs and sperms are single cells which cannot survive long in sea water; unless an egg is fertilized within a short time, it perishes. Sperms are attracted to the egg, and the first one that penetrates the membrane (which it does rapidly, leaving its flagellum outside) starts a physicochemical change such that a large amount of fluid collects beneath the membrane, elevating it and serving as at least one of the factors which prevent other sperms from entering. The sperm nucleus migrates towards the egg nucleus, and both undergo changes preparatory to the first cell division, which is initiated by the entrance of the sperm. Each nucleus brings the hereditary contributions of one parent to the new individual.

By subjecting unfertilized eggs to various stimuli, such as certain acids or concentrated sea water, it is possible to activate the egg and cause it to divide, thus duplicating the effect of the sperm. The larva which develops is normal in every way, though, of course, it has no paternal characteristics. Strong centrifuging of unfertilized eggs causes them to become dumb-bell-shaped and finally to break in two. The lighter halves contain the maternal nuclei; the heavier, non-nucleated halves can be collected and fertilized by adding sperms to the dish. The larvas thus obtained have a father but no mother. Such methods make it possible to isolate maternal and paternal characteristics and also to prove that the reproductive cells of both sexes make equivalent contributions to the heredity of the offspring. Finally, fertilized eggs can be centrifuged strongly until they break in two. The nucleated fragment develops normally. The non-nucleated fragment divides several times but eventually dies (like the *Euplotes* from which the small nucleus was removed; see p. 36, Vol. 1). From these experiments we can see that either a male or a female nucleus is sufficient for development, but apparently at least one nucleus is necessary for continued differentiation and life.

After fertilization the zygote divides into two equal cells; and these divide into four, eight, sixteen, and so on until there results

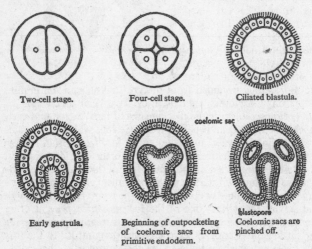

Two-cell stage.   Four-cell stage.   Ciliated blastula.

Early gastrula.   Beginning of outpocketing   Coelomic sacs are
                    of coelomic sacs from   pinched off.
                    primitive endoderm.

EARLY DEVELOPMENT OF A STARFISH. (After Delage and Herouard.)

a hollow BLASTULA, which is ciliated and swims about. By the end of the first day after hatching from the egg, the free-swimming blastula has been transformed, by an infolding of the cells at one pole, into a two-layered GASTRULA with an outer ectoderm and an inner endoderm. This infolding crowds out most of the old BLASTULA CAVITY (blastocoel) and produces the PRIMITIVE GUT CAVITY (archenteron), which will later become the cavity of the digestive tract. The opening into this new cavity is called the BLASTOPORE. Through the gastrula stage the embryos of most animals are essentially alike, although they differ in size, in rate of development, in amount of yolk, in the equality of division of the cells, and in the particular way in which the gastrula becomes two-layered. As development proceeds, the number of differences increases. In the starfish the next stage is characterized externally by the elongation of the larva, a decrease of cilia over the general surface, and an increase in the cilia along definite bands. In the meantime important internal changes take place. From the endoderm arise two out-pocketings, which gradually pinch off as two coelomic sacs. These eventually give

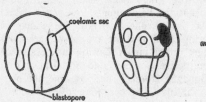

Coelomic sacs divide into anterior and posterior parts.

Left anterior coelom sends outgrowth to dorsal surface; this is forerunner of stone canal of adult.

Left anterior coelom begins to curve around mouth and produce buds. The larval anus is derived from the blastopore.

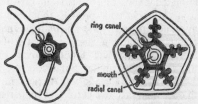

Rudiments of five radial canals are seen as buds from the left anterior coelomic sac.

The right and left posterior sacs form the main body cavity of adult.

A new anus has formed dorsally and does not show in this view of the ventral surface.

FURTHER DEVELOPMENT OF A STARFISH. (After Delage and Herouard.)

rise to the COELOM and its derivative, the water vascular system. Up to this point the larva has been bilaterally symmetrical. But after each coelomic sac divides into an anterior and a posterior portion, changes begin which transform the bilateral larva into a radial adult. The left anterior coelomic sac sends a tubular outgrowth to the upper surface, where it opens by a pore in the ectoderm. Cilia lining this tube draw water into the coelom, and this pore canal is the forerunner of the stone canal of the adult. In some starfishes the bilateral symmetry persists some-what longer, and the right anterior coelom also sends a pore canal to the upper surface, though this canal finally closes up and is lost. This suggests that the ancestral starfish originally pro-duced similar organs on both sides, and that the radial con-dition of modern starfishes is a result of the degeneration of the right-hand side of the embryo and an over-growth of the left side. The hind portion of the left anterior coelom soon produces another hollow outgrowth, which encircles the mouth and be-comes five-lobed. This is the beginning of the water vascular system (shown in solid black in the diagram). The circular out-growth becomes the ring canal, and the five lobes represent the rudiments of the future radial canals. During the development of the coelom and its derivatives the endodermal layer differ-entiates into the oesophagus, stomach, and intestine; the blasto-pore serves as the larval anus. The oesophagus bends ventrally and meets a tubular ingrowth from the ectoderm, which breaks through to form the mouth. Meanwhile, the external surface of the larva has been greatly expanded by the extension of lobe-like processes which enable the larva to float easily. Certain of the ciliated bands are carried out on these processes, and the beating of their cilia propels the larva about. Food-bearing currents are created by the cilia on bands near the mouth.

The free-swimming starfish larva (called a *bipinnaria*) has bilaterally symmetrical lobes bearing ciliated bands. Other echinoderms have similar but distinctive larvas, some of them with very much longer lobes. At least in the earlier stages, all echinoderm larvas have certain characteristics in common. They are bilaterally symmetrical, swim by means of longitudinal looped ciliated bands, have a similar digestive tract and a coelom which connects with the upper surface by a tubular canal. All this suggests that there must have been some ancestral echino-derm larva which embodied all these features and from which all the different classes of echinoderms have been derived. This hypothetical ancestral larval type (called the *dipleurula*) contrasts strongly with the trochophore-like ancestor from which, it is believed, has come the flatworm-mollusc-annelid stock.

To return to the starfish development, we see next that the bilateral larva settles down on some solid object and remains temporarily fixed while it becomes transformed into the adult starfish. The left and right posterior coelomic sacs form the general body cavity of the adult. The larval mouth and anus close, and a new mouth breaks through on the original left side, while a new anus opens on the original right side, thus producing an adult axis at right angles to the larval axis. The radial canals grow out and develop tube feet. And the first signs of the adult body form appear as five elevations of the ectoderm.

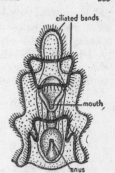

The free-swimming LARVA OF A STARFISH is bilaterally symmetrical and has ciliated bands for locomotion and feeding. (After Field.)

The symmetry of the adult starfish is called SECONDARY RADIAL SYMMETRY, because we believe that it is only secondarily derived from an originally bilateral ancestor. This is suggested in the adult by the asymmetrical position of the sieve plate. But the most important evidence comes from the larval development in which, by a process of asymmetrical growth, the bilateral larva is converted into a radial adult.

The fact that the bilateral starfish larva becomes temporarily attached during the time that it changes over into the radial adult is very interesting in view of what we believe to be the ancestral history of echinoderms. The earliest-known echinoderm fossils are fixed types. And some of the modern sea lilies (a class of echinoderms) have a free-swimming bilateral larva which becomes attached and then metamorphoses into a permanently fixed adult. As we have seen before, bilateral symmetry is the symmetry of fast-moving animals and radial symmetry seems best suited to sessile animals, which must meet their environment on all sides. It is considered likely, therefore, that the ancestral echinoderm was a bilateral free-living animal which became radial and took up sessile habits secondarily, and that the modern free-living echinoderms are derived from a fixed ancestor, whose symmetry they still retain.

SEA URCHINS seem very unlike starfishes; yet they have the same fundamental structure. A sea urchin looks like a huge animated burr with long sharp spines that are movable and aid the tube feet in locomotion. Instead of numerous small stony pieces embedded in a muscular wall, the sea urchin has plates

which are fused to form a rounded shell completely inclosing the
soft parts. In the centre of the lower surface of the shell is an
opening for the mouth, and the anus opens by a small hole on
top. Radiating upward from mouth to anus are five bands of
minute holes through which the tube feet project. These five rows
correspond to the rows of tube feet on the five arms of the star-

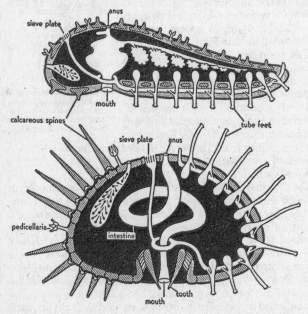

The sea urchin is much like the starfish in basic plan. *Above*, a longitudinal section
through the disc and one arm of a starfish. *Below*, a section through a sea urchin.

fish. If we imagine the arms of the starfish bending upward to
meet, and if, at the same time, we fill in the angles between them
by elevating and reducing the size of the disc so that the sieve
plate will lie at the ends of the rows of tube feet, we can see how
the round sea urchin is similar to the five-rayed starfish.

The sea urchin has the same type of water vascular system that
was described for the starfish, but the feet are more slender and
much longer, since they must project beyond the spines. The
pedicellarias have three jaws and are usually on long stalks. The

digestive tract is longer than that of the starfish, and the sea urchin feeds mostly upon vegetation, which requires more prolonged digestion. Around the mouth is an elaborate set of five teeth, arranged radially and worked by muscles in such a way as to chew the food. The stomach cannot be turned inside out, and the intestine is long and coiled. The anus is functional and is located to one side of the sieve plate. Other systems are very much like those described for the starfish. The gonads open from five pores near the sieve plate.

Not all echinoderms have prominent spines. The sea cucumbers have a leathery skin in which are embedded microscopic plates of calcium carbonate. The plates shown here were isolated by mounting a fragment of skin on a microscopic slide and dissolving away the living tissue.

THE other echinoderm groups – brittle stars, sea cucumbers, and sea lilies – all deviate from the starfish about as much as the sea urchin but have characteristically a spiny or leathery skin, a water vascular system based upon a plan of five or multiples of five, an extensive coelom, and a bilateral larva. These groups are illustrated by photographs.

CHAPTER 26

Invertebrate Chordates

THE only major group we have not yet considered is the phylum CHORDATA, composed almost entirely of animals *with* backbones and therefore no proper subject for this book. There are, however, a few 'lower chordates' which have no vertebral column. These are mostly inconspicuous animals, seldom seen, or at least not usually noticed. None is of much economic importance, though the amphioxus is so abundant on the seacoast of China that during certain months these small animals are collected by the ton for human consumption.

The amphioxus, and the other invertebrate chordates, the tunicates and acorn worms, are interesting chiefly because they share with the vertebrates certain distinctive characters found nowhere else in the animal kingdom, and so serve to link the vertebrates with the invertebrates.

These chordate characters are all well shown by the AMPHI-OXUS ('sharp at both ends'), a small, laterally compressed, semi-transparent animal which lives in shallow marine waters all over the world. It can swim about by fishlike undulations of the body, but spends most of the time buried in the sandy bottom with only the anterior end protruding above the surface. In this position it feeds by drawing into the mouth a steady current of water, from which it strains suspended microscopic organisms.

The most distinctive chordate character, and the one from which the name of the phylum is derived, is the NOTOCHORD, a cartilage-like rod which extends the length of the body and gives support to the soft tissues. It serves as a rigid but flexible axis on which the soft muscles can pull, and so permits powerful side-to-side undulatory movements of the whole body, which carry the animal through the water with a speed unattainable by flabby animals like the planaria or the nereis.

The strong, swift swimming movements of aquatic chordates, made possible by a flexible internal support for the muscles, were probably a major factor in the early success of the group.

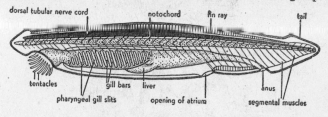

AMPHIOXUS.

Besides great muscularity (as shown in the cross-section of the amphioxus), chordates are characterized by the prolongation of the body beyond the anus as a *tail*, a region specialized for swimming and containing little else but the skeletal axis, nerves, and muscles. In the amphioxus the tail is very small, and there are no paired fins as in fishes; but running along the entire dorsal surface, and extending around the posterior end on to the ventral surface of the body for a short distance, is a ridge supported by a series of gelatinous *fin-rays*.

The notochord is derived from the roof of the primitive endoderm and is present in the embryos of all chordates including man. In primitive vertebrates (lampreys) and some fishes the notochord persists, and the backbone, which is composed of a series of hard separate vertebras and is stronger though just as flexible, forms around it. In all higher vertebrates the backbone replaces the notochord as the mechanical axis of the body.

A second chordate character is the DORSAL TUBULAR NERVE CORD. In all other invertebrates which have been described, the principal nerve cord was ventral or lateral in position; but in the amphioxus and other chordates, it lies between the notochord and the dorsal skin and is hollow. From the cord go a pair of nerves to each of the segmentally arranged bundles of muscles.

The third important chordate character is the structure of the pharynx, which is perforated by pairs of slitlike openings. The

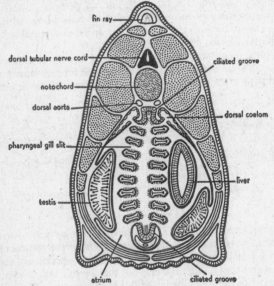

CROSS-SECTION THROUGH AN AMPHIOXUS.

PHARYNGEAL GILL SLITS, so named because they have mainly a respiratory function in fishes, in the amphioxus serve chiefly as an apparatus for straining food from the water. The pharynx is lined with cilia, which beat inward, producing a steady current of water that enters the mouth and passes out through the pharyngeal slits, leaving behind the suspended particles. The slits do not open directly to the exterior, as in fishes, but into a chamber, the *atrium*, which surrounds the pharynx and opens to the exterior by a pore some distance anterior to the anus. The atrium is lined with ectoderm and is formed, in the embryo, by the outgrowth of two folds of skin, one on each side, which finally fuse in the mid-line. The walls of the pharynx, being perforated from top to bottom by the vertical gill slits, would collapse if it were not for a supporting framework of rods which run in the walls bounding the gill slits. The tissue with its rod, which lies between any two gill slits, is called a *gill bar*.

The amphioxus embryo has about sixty pairs of gill slits (and they increase in number as the animal grows). This large and indefinite number is a primitive condition; and in the shark, for example, there are always six gill slits. Land vertebrates, which breathe by lungs, have no gill slits in the adult stage, but the slits make a fleeting appearance in the embryo. The human embryo develops gill pouches, but slits never break through.

Except for the gill slits, most of the structures associated with the feeding mechanism of the amphioxus are peculiar to the animal and must not be thought of as primitive chordate characters. At the anterior end of the animal is a funnel-like hood fringed with a row of sensory tentacles. At the back of the hood is the *mouth*, a circular aperture bounded by a rim of small tentacles which can be brought together to form a kind of strainer for keeping out large particles. The cilia in the pharynx beat inward and downward, so that water drawn into the mouth is directed toward a *ciliated groove* in the floor of the pharynx, where the suspended organisms are trapped in mucus. By the action of the cilia the food-laden mucus is moved forward to the anterior end of the pharynx, upwards on each side, and then backwards along a ciliated groove in the roof of the pharynx, to the *intestine*. There the food and the mucus are digested and almost completely absorbed. From the anterior end of the intestine is given off a hollow gland, the *liver*, which extends forward along the right side of the pharynx and can be seen in the cross-section. This gland secretes a digestive fluid; and since it arises in the same way as the vertebrate liver, by an outpocketing of the digestive tract, the two organs are thought to be homologous. Further, the blood leaving the capillaries of the intestine of amphioxus is not returned to the general circulation until it has passed through the capillaries of the liver. Such a path for the blood is found nowhere else except in vertebrates, and it furnishes striking evidence that the amphioxus is descended from the same primitive stock which gave rise to the backboned animals.

The *circulatory system* is a closed one; and there is no heart, the blood being pumped by the contractile ventral vessel. In this connection it is interesting to note that in all vertebrate embryos the blood is first pumped by a simple, pulsating tube which only later becomes bent and constricted to form the heart. The blood receives oxygen as it flows through vessels in the gill bars, which are in close contact with the steady current of water passing through the pharynx. After passing through the gill bars, the blood flows into two vessels, the dorsal aortas, which unite behind the pharynx into a single vessel that supplies the intestine. Wastes are extracted from the blood by *excretory organs*, a pair to each pair of gill slits, which lie against the dorsal wall of the pharynx and resemble, strangely enough, the excretory organs of annelids. The *coelom* of the amphioxus has been partly crowded out by the expansion of the atrium. In the pharyngeal region it is represented only by two small cavities, which lie, one on each side

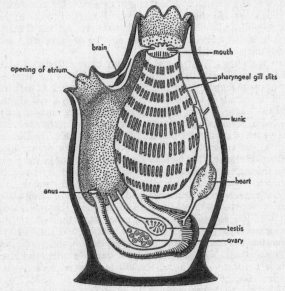

TUNICATE. (Modified after Delage and Herouard.)

of the pharynx, above the atrium. The sexes are separate; and the *reproductive organs* are paired, segmentally repeated pouches which lie along the sides of the body and push the atrial walls inward, so that in the cross-section they appear to lie in the cavity of the atrium.

There are no paired eyes or other well-developed sense organs, though there is a large pigment spot at the anterior tip of the nerve cord and a row of smaller pigment spots along the lower edge of the cord. The nerve cord does not expand at the anterior end into a brain but is tapered to a point. The apparent simplicity of the sense organs and central nervous system is probably not a primitive condition, but is, more likely, a degeneration of the head region associated with the sedentary habits of the animal.

In spite of its poor nervous system and various specializations associated with its particular way of life, the amphioxus is the only living form to which we can look for a concrete idea of what the primitive chordates, which gave rise to the vertebrates, might have been like.

THE TUNICATES are so named because the outer layer of the body wall is a tough, often translucent 'tunic' made of a material very much like the cellulose of plants. Some tunicates are free-living and swim near the surface; but most forms are sessile,

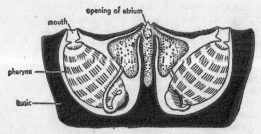

Two members of a TUNICATE COLONY. They are embedded in a common tunic and share the same atrial opening, but have separate mouths. (Modified after Delage and Herouard.)

growing permanently attached to rocks or seaweeds. The simplest kind looks like an upright sac with two openings: one at the top, and one somewhat lower down on one side. When the animal is disturbed, the body wall contracts suddenly, and the water contained within the body is forced out in two jets – hence the common name 'sea squirt'. The interior of the body is occupied for the most part by a large saclike pharynx perforated by many rows of *pharyngeal slits*. Cilia round the edges of the slits create a current which draws water into the mouth at the top, through the pharyngeal slits, and out into the atrium, a cavity surrounding the pharynx. Food particles are trapped in the pharynx, and water leaves by an opening from the atrium.

The atrium also serves as an exit for faeces and sex cells, since the anus and sex organs (the animal is hermaphroditic) open into the atrium. There are few real blood vessels, the blood flowing mostly through

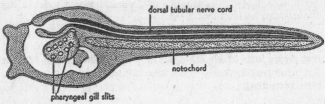

The free-swimming LARVA OF A TUNICATE has all three chordate characters. (Based on several sources.)

spaces among the tissues. The heart is remarkable in that it drives the blood in one direction during several beats and then reverses the direction.

Many tunicates reproduce asexually, as well as sexually. The buds fail to separate and the individuals remain together as a colony, embedded in a common tunic, which, in the case of sessile tunicates, grows as an encrusting mass over the surface of rocks, shells, or sea weeds. The members may be arranged in small groups; and in the colonial tunicate, shown in the drawing which heads this chapter, there are four such groups represented. Each is star-shaped, and at the tips of the rays are the separate mouths of the several members. At the centre of the star is a common opening for the exit of water.

Except for the gill slits there is very little about such a simple, sessile animal to suggest any reason for including it in the same phylum with fishes or mammals. But the development of a tunicate tells another story. The larva is a free-swimming animal that reminds one of a tadpole. It has a large tail

ACORN WORM. (After Bateson.)

which contains, besides muscles, a well-developed *notochord* and a *dorsal tubular nerve cord*. The trunk contains, in addition to other organs, a *pharynx with slits*. The larva finally settles down on a rock, loses the tail containing the notochord and nerve cord, and revamps its whole structure in adaptation to its sessile life. The adult has no trace of a notochord, and the nervous system is represented only by a ganglion in the dorsal region of the body between the two openings. Here we have another striking example of how animal relationships can be established through a study of the young stages, even though the adult is a degenerate form whose similarities to other animals are quite obscure.

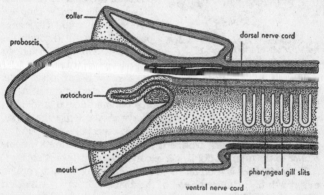

Longitudinal section through ANTERIOR END OF ACORN WORM. (Based on several sources.)

THE most primitive of the invertebrate chordates are the ACORN WORMS, soft, elongate animals which burrow in the sand or mud of seashores. At the anterior end of the body is a muscular *proboscis* joined by a narrow stalk to a short, wide *collar*. The worm shown in the accompanying drawing has a long cylindrical proboscis, but in many forms (as in those in the drawing at the head of the chapter) the proboscis is ovoid. Its shape, and the way it fits into the collar when not extended, reminded someone of an acorn – hence the name of the worm. The proboscis and collar are muscular and are used in burrowing. The proboscis is forced through the sand, with the collar following. The distension of the collar firmly anchors the anterior end of the worm,

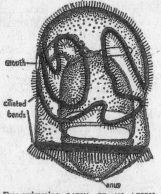

so that contraction of the muscles in the trunk region draws the trunk forward.

The mouth is in the middle of the ventral surface, at the base of the proboscis and concealed by the edge of the collar. As the animal burrows, sand or mud passes into the mouth, through the digestive tract, and out the anus at the posterior tip. Organic materials present in the sand or mud are digested. Water taken in is supposed to pass out through the *pharyngeal gill slits* which open through the dorsal wall of the anterior part of the trunk.

Free-swimming LARVA OF AN ACORN WORM. (Modified after Morgan.)

The nervous system consists of a network of nerve cells extending under the whole of the surface ectoderm and in the trunk region is concentrated along the mid-dorsal and mid-ventral lines of the body as two nerve cords. Only the *dorsal cords* extend into the collar, where it is especially thick, and in some species of acorn worms it is hollow, suggesting a resemblance to the tubular, dorsal nerve cord of a typical chordate.

The *notochord* is thought to be represented by a short rodlike outgrowth of the anterior end of the digestive tract into the base of the proboscis. This outgrowth strengthens the proboscis and is composed of vacuolated cells like those in the notochord of the amphioxus, but whether it really corresponds to the notochord of typical chordates is not clear.

Apart from their interest as animals which appear to have branched from the early chordate stock, the acorn worms, through their larvas, furnish one of the few real clues that link the chordates with any other phylum. The larva of the acorn worm looks so surprisingly like the larvas of certain echinoderms that they can be mistaken for them. Moreover, similarities extend beyond the structure of the two organisms and include several details in the formation of the coelom and other parts. The later development of the larvas is very different, for the echinoderm larva metamorphoses into an adult with a kind of radial symmetry, while the adult acorn worm is bilateral.

MEMBERS of the phylum Chordata all have, at some time in their life-history, a notochord, dorsal tubular nervous system, and pharyngeal gill slits. But in most other respects they fall into four groups – vertebrates, amphioxus, tunicates, and acorn worms – so different that each is designated a subphylum. Of the three groups of invertebrate chordates, the amphioxus is most, and the acorn worm least, like the vertebrates. Since acorn worms have only a poor excuse for a notochord and dorsal tubular nerve cord, some zoologists question whether they deserve the status of chordates at all, and would place these wormlike animals with gill slits in a phylum by themselves.

CHAPTER 27

Records of the Invertebrate Past

ALTHOUGH animals have left none of the birth certificates, marriage contracts, tombstone inscriptions, or written documents, upon which students of human history depend so much, there are abundant records of the invertebrate past – not just of the past few thousand years but of some 550,000,000 years or more.

Any evidence in the materials or rocks of the earth's crust that gives some idea of the size, shape, or structure of the whole or any part of a plant or animal that once lived is called a FOSSIL. The name is derived from the Latin 'to dig' because it was, at one time, applied to almost anything of interest that was dug up. Coal is an indication of past life, but it cannot be considered a fossil because by itself it gives no idea of the organisms which were responsible for its formation. Nor would we classify as a fossil any empty snail shell that turned up while we were digging about in the garden, for not only must a fossil have something of the character of an organism but it must have age. The exact age at which animal remains become fossils is not fixed; most fossils are evidences of organisms that lived at least 25,000 years ago; and most, though not all of them, are of species now extinct.

Fossils can be formed in a variety of ways. The rarest, but also the most interesting, kinds are ANIMALS PRESERVED WITH LITTLE OR NO CHANGE FROM THEIR CONDITION AT THE TIME OF DEATH. Such are found in considerable numbers in the amber of the Baltic region of Europe, which in early Oligocene times (about 38,000,000 years ago) was covered with a dense coniferous forest. Drops of gummy resin, dripping from the trees, trapped spiders and mites scurrying across the forest floor or insects on the wing. The sticky resin, on fossilization, became hard amber; and the arthropods embedded in it are preserved intact, sometimes even with their colours unchanged. So perfectly embalmed are some of the specimens that they have been dissected and their intestinal parasites examined. Less complete, but much more common, unaltered fossils are the shells of molluscs and brachiopods, which, even after millions of years of preservation in the rocks, may still retain their colours or lustre.

Most fossils, however, have undergone some change since the death of the organism. Horny coverings, such as the chitinous exoskeletons of arthropods, leave only thin films of carbon in the rocks. Calcareous shells and other skeletal structures are slowly dissolved by water percolating through the ground and are gradually replaced by minerals, such as calcite, silicon dioxide, or iron sulphide, which are deposited in the cavities left by the slow dissolution of the original materials. Shells may be completely dissolved away, leaving only a cavity, on the wall of which is a MOULD of the external surface of the shell. Or the cavity later may be filled with some mineral, forming a CAST of the original fossil.

As we would expect, the vast majority of animal fossils are of forms with hard parts: calcareous shells or spicules; horny coverings, exoskeletons, or jaws; and silicious coverings or spicules. But even very soft animals like jellyfishes may leave IMPRESSIONS on soft mud which, if soon covered by a layer of fine sediment, may eventually be preserved when the mud hardens into rock. In a similar way there have been formed many of the INDIRECT EVIDENCES of the former activities of animals. *Tracks or trails* impressed on muddy or sandy bottoms sometimes indicate the kind of appendage that left the record. *Burrows* in mud or sand and *borings* in rocks can be identified if, as in the case of sea urchins and pelecypods, the shells or parts of the shells are left in the cavities. Certain fossil markings, either straplike in form or resembling the pellets eliminated by living animals, are interpreted as fossilized animal excreta. In some cases these *coprolites*, as they are called, have revealed what the extinct animals fed on.

In order to be fossilized, an animal must be buried at the time of death or very shortly thereafter; otherwise the body is likely to be eaten by scavengers or disintegrated by bacteria. FOSSILS FORMED ON LAND are relatively rare, for, even when land animals are covered by wind-blown earth or sand rapidly enough to escape destruction by scavengers or bacterial decay, the potential fossils are soon removed by erosion. Only under exceptional conditions, such as prevail when an eruption traps animals under heavy layers of volcanic ash, or dripping resin accumulates over long periods, or land animals are caught in floods and are carried out to sea, do we get fossils good enough to contribute to our sketchy picture of ancient terrestrial life. FOSSILS OF MARINE INVERTEBRATES, however, are very abundant, for rapid burial is easily accomplished in shallow sea water, where there is constant shifting of the mud or sand at the bottom. Later, when by geological processes these marine sediments have been uplifted from the sea bottom, they present a legible record, with the earliest animals preserved in the lowermost layers and the most recent types in the uppermost strata. In most places we can examine only the layers fairly close to the surface, and we have no access to the hundreds of millions of fossils in the underlying rocks. Over certain large areas, however, the older rocks have been exposed by uplift and erosion. They can also be examined in places like the walls of the Grand Canyon of the Colorado, where the river has cut a cross-section a mile deep through the earth's crust.

A systematic study of the fossils found in the successive layers of rock reveals, not only that animals have been present on the earth during at least 550,000,000 years for which we have good fossil records, but that *the deeper we go into the rocks the less and less familiar are the fossils which we find*. Those a mere million years old are of animals much like living forms. Those several million years old are more different, and those still older belong to orders, and even to whole classes, of animals which no longer exist. In other words, the fossil record furnishes direct evidence that animals were not always as they are today, and that modern forms, which do not occur as fossils, must be descended, by a process of gradual modification which we call *organic evolution*, from the earlier and simpler animals whose remains we find in the rocks. In some cases the record is so complete that we are able to trace the gradual evolution, from one layer of rock to the next, of a definite structure or set of structures, all intermediate stages from the most primitive type to the modern living form being present.

Strangely enough, the science of ancient life, or paleontology, has been developed not by biologists, who have been busy enough studying living animals, but by geologists, who have demonstrated that the rocks of the various periods in geologic time are characterized by a distinctive assemblage of animal and plant fossils. Certain of these fossils, which are world-wide in distribution, distinctive, and restricted to limited periods in geologic time, serve as markers or *index fossils* by which rocks of any period can be recognized no matter where they occur.

GEOLOGICAL time has been divided into six ERAS, the end of each marking the time of some significant geological change, such as continental uplift and mountain-building. Shorter and less profound changes in the face of the earth form a basis for dividing each of the eras into a number of PERIODS.

The first, or AZOIC ('no life'), era marks the origin of the earth from the sun and the formation of rocks; no life was present. This period lasted over 1,000,000,000 years (a time estimate based on several lines of evidence, mostly on the rate of disintegration of radioactive substances). In the second, or ARCHEOZOIC ('primitive life') era, there was extensive volcanic activity and mountain-building. If life had evolved, the rocks show little evidence of it, though they do contain carbonaceous material, probably an indication of primitive algae. Perhaps one-celled animals appeared at this time. Rocks of the PROTEROZOIC ('first life') era have only rarely yielded a recognizable fossil; yet this era must certainly have been a time of great evolutionary development, for by the *Cambrian* period, the first period of the PALEOZOIC ('ancient life') era, the animal kingdom is already highly diversified. Nearly all phyla which leave any kind of a fossil record are well represented in Cambrian rocks – many of them by several groups, which already show the distinctive characters of modern classes. Why pre-Cambrian fossils are so rare is not yet understood, though many possible explanations have been suggested. Two of the most plausible ones are that animals did not evolve hard, preservable parts until the Cambrian and that the pre-Cambrian rocks, being very ancient, have been subjected to so many stresses that any fossils which they once contained have been destroyed. In any case, the earliest well-preserved, abundant fossils are those laid down in the Cambrian period; and from that time on they increase steadily with the expansion of the various groups. By the *Ordovician*, the second period of the Paleozoic, many invertebrate groups are at their peak of abundance, and vertebrates are already on the scene. At the close of the Paleozoic, many of its most important invertebrate groups become extinct. The MESOZOIC ('middle life') is an era of waning influence for many marine invertebrate groups, though certain large, shelled cephalopods reach their climax during this time. At the beginning of the CENOZOIC ('recent life') era there appear many of the more modern types of invertebrates.

THE major time units, their estimated age and length of duration, together with some of the invertebrate groups characteristic of the various periods, are summarized in the accompanying table. The rest of this chapter is devoted to a brief description of some of the better-known fossil groups.

AMONG the PROTOZOA one would naturally not expect to find fossils of flagellates, amoebas, or ciliates, but only of the amoeboid protozoans with hard parts. Although the silicious skeletons of *radiolarians* are known from the Paleozoic to the present, the calcareous shells of the *foraminifers* furnish by far the best record of ancient protozoan life. Good fossil foraminifers occur in the Cambrian rocks, increase steadily both in numbers of species and of individuals, until in the Cretaceous they finally become very abundant; and there appear many families which are still represented today. From the Tertiary to the present there has been a gradual extinction of the earlier types and a replacement by those more like modern foraminifers. This gradual change in type, plus the fact that the calcareous shells are abundant, widespread, easily preserved, and have distinctive shapes and markings, makes fossil foraminifers extremely valuable as index fossils. There are, of course, many different kinds of index fossils; but most of them, like trilobites and brachiopods, are large and can be obtained only from rocks which are exposed.

TABLE OF GEOLOGIC TIME

Eras (Millions of Years Ago*)	Periods (Duration in Millions of Years*)	Principal Evolutionary Events among the Invertebrates
CENOZOIC (60)	Quaternary (2)	Arthropods and molluscs most abundant; all other phyla well represented
	Tertiary (58)	Modern invertebrate types appear
MESOZOIC (200)	Cretaceous (70)	Extinction of ammonoids
	Jurassic (38)	Ammonoids abundant; modern types of crustaceans appear
	Triassic (32)	Marine invertebrates decline in numbers and importance. *Limulus* present
PALEOZOIC (550)	Permian (35)	Last of the trilobites and eurypterids. Extinction of most paleozoic invertebrate types
	Pennsylvanian (45)	First fossil insects (although insects probably evolved as early as the Devonian)
	Mississippian (35)	Climax of crinoids and blastoids
	Devonian (35)	Brachiopods still flourishing. Marked decline of graptolites and trilobites
	Silurian (25)	Many graptolites. First extensive coral reefs. Brachiopods at peak. Eurypterids abundant. Trilobites begin to decline. First land invertebrates
	Ordovician (70)	Peak of invertebrate dominance. Climax of trilobites and nautiloid cephalopods. Brachiopods abundant. (First vertebrate fossils)
	Cambrian (105)	First abundant fossils. Nearly all invertebrate phyla represented. Trilobites and brachiopods most numerous
PROTEROZOIC (1,200)	(650)	Most of the invertebrate phyla probably evolved, but fossils are rare and poorly preserved
ARCHEOZOIC (2,000)	(800)	No fossil record, but simplest living organisms (one-celled plants and animals) probably arose
AZOIC (3,000)	(1,000)	No evidence of living organisms. (Formation of the earth)

* Based on Croneis and Krumbein, *Down to Earth*.

Because of their minute size, foraminiferan shells can be obtained undamaged, from rocks very far below the surface, in the borings brought up by a drill. For this reason oil companies have found it profitable to employ paleontologists who do nothing but study the borings from oil-well drills. By comparing the shells found at various levels with those present in layers known to be oil-bearing, they are able to direct the well-drilling operations.

THE fossil record of SPONGES is not an abundant one, but it stretches over a very long time, silicious spicules being known even from the pre-Cambrian, and whole specimens from the Cambrian. The fibres of horny sponges have practically no chance of being preserved, and the delicate spicules of calcareous sponges are usually dissolved under the local acid conditions created by the decay of a dead sponge. Silicious spicules are more readily preserved but may later be dissolved away by ground water percolating through the rocks, and in such cases the cavities left may be filled with calcareous or other materials. Consequently, the composition of a sponge fossil does not necessarily indicate the original composition of the living sponge, and identification must be based on the general structure and on the shape of the spicules.

FOSSILS of COELENTERATES are known from the pre-Cambrian to the present. Although pre-Cambrian fossils of any kind are extremely rare, and after many years of search no other well-authenticated fossil has ever been found in the pre-Cambrian rocks exposed in the Grand Canyon, those rocks have yielded a well-preserved imprint of one of the most fragile of animals, a jellyfish. Finds like this one serve to emphasize the fortuitous nature of any fossil record.

The HYDROZOA are represented in the lower Cambrian layers; and in the upper Cambrian rocks we find the beginnings of the GRAPTOZOA, an interesting group of colonial animals, all extinct, that have been claimed for both the sponges and bryozoans but are usually placed among the coelenterates. Graptolites are most often found as thin carbonaceous films which look like saw blades with one or both edges toothed. A blade is the remains of a horny colonial skeleton, and each notch in the toothed edge represents a small cup which formerly housed a single member of the colony. From the blade there comes off a long thread by which, it is thought, the colony was attached to floating sea-weed. Some graptolites are found in large groups radiating from some central body, which may have been a gas-filled float. If so, they were independent floating colonies comparable with our modern siphonophores. The types just described flourished in the Ordovician and Silurian and died out before the beginning of the Devonian – possibly because of something related to the rise of the fishes. The more primitive branching graptolites, most of which are thought to have lived attached to the bottom, appeared in the Cambrian and had a few representatives which managed to survive throughout the Devonian. This situation is what we find in most groups; the primitive forms not only precede the more specialized and 'progressive' ones, but, being less dependent upon special conditions, usually also outlast them.

Fossil SCYPHOZOA are known from the lower Cambrian, but they are rare and the authenticity of some is questioned. They consist of what appear to be moulds of the upper and lower sur faces and mud fillings of the pouches which comprise the large central part of the gastrovascular cavity.

The ANTHOZOA, as one might predict from their highly developed structure, are the latest class to appear in the fossil record. The sea anemones, lacking hard parts, have left no recognizable traces. But the forms which secrete skeletons, particularly those made of limestone, provide an abundant and informative record of the past activities of their delicate and perishable polyps. Two classes of extinct corals are known. One of these, the *Tetracoralla* (named from the fact that the stony partitions are in multiples of four), occurs as solitary cup corals or as dome-shaped or branching colonies. The skeletons appear in the Ordovician, attain their maximum in the Silurian, when certain ones contribute abundantly to the fossil reefs, and disappear by the end of the Paleozoic. The *Hexacoralla*, to which modern anemones and stony corals belong, are not represented until after the beginning of the Mesozoic. Their sudden, though late, appearance is accounted for by assuming that they are descended directly from the exclusively Paleozoic Tetracoralla. The stony corals were important contributors to the Mesozoic and Cenozoic reefs and are the dominant reef-builders of modern seas. The *Octocoralla*, or alcyonarians, anthozoans with eight tentacles and eight internal partitions, are represented from the Ordovician on.

THE wormlike invertebrates have left a very poor fossil record. They may be discussed together because part of the record consists of trails, tracks, tubes, and burrows; and it is impossible to say which of the various groups is responsible for certain of them. The FLATWORMS have left no recognizable remains. Fossil NEMERTEANS are unknown; but, since the living forms do burrow, it is possible that some of the fossil trails and burrows were made by members of this group. Fossil parasitic ROUNDWORMS are reported to have been found in fossil insects.

OF the ANNELIDS, only the polychetes have left a record of much importance. From the pre-Cambrian we have tracks and burrows, many of which must have been left by this group. From the mid-Cambrian we have well-preserved fossils with segmental bundles of bristles and other characteristic features of the class, showing that the animals must have had a long pre-Cambrian development. The free-swimming polychetes, modern representatives of which have a protrusible pharynx armed with hard jaws, are thought to be the source of the numerous small-toothed jaws found in practically all rocks from the Paleozoic on. Beginning with the Ordovician, we find evidence of tube-dwelling polychetes in the form of tubes which are calcareous or are made of sand grains or other particles cemented together by a secretion. Some of them are found attached to brachiopod shells, often in clusters.

THE BRYOZOA appear first in the Ordovician, and apparently have been quite abundant from that time to the present, for over 1,500 species are known from the Paleozoic, 1,000 from the Mesozoic, and there are additional thousands of Cenozoic and modern species. While whole colonies are sizable (a number of fragments of the skeletons of branching colonies appear in the drawing which heads this chapter, and in the drawing look like pieces of broken twigs), the individual cases are microscopic and therefore have many of the same advantages for the paleontologist as foraminiferan shells. Two whole classes of Paleozoic bryozoans became extinct at the end of that era. One class, already represented in the Ordovician, and which has never been very abundant, still has living representatives. The class to which most modern bryozoans belong did not appear until the Jurassic.

WE are accustomed to think of the BRACHIOPODS as a small and unimportant phylum, which in this book has been described as one of the 'lesser lights'. But if we are to view the animal kingdom in terms of the past as well as the present, we would have to reserve a more prominent place for the group that has left one of the most abundant, most complete, and most beautifully preserved of all fossil records. Not only do brachiopods have hard shells which lend themselves readily to preservation, but the animals live in shallow seas where the chance of fossilization is great. Consequently, their remains are often so abundant as to form most of the rock in which they occur. The group left its earliest record in the Cambrian; and before the end of that period four of the five orders were already established, with the dominant position occupied by the relatively primitive order still represented by *Lingula*. Brachiopods with hinged valves were present in the Cambrian but did not become of major importance until the beginning of the Ordovician. Something of the early abundance and later decline of the brachiopods can be judged from the numbers of fossil genera recorded for the several eras. We know about 450 genera from the Paleozoic, about 180 from the Mesozoic, and only about 75 from the Cenozoic to the present. Although 3,000 species are recorded from the Ordovician and Silurian periods, which represent the peak of brachiopod abundance, only a little over 200 species are living today.

BEING shelled animals and second only to the arthropods in numbers of species, the MOLLUSCS have left an abundant, unbroken, and extremely legible record from the Cambrian to the present. The *amphineurans* are certainly the most primitive of modern molluscs, but they do not appear as fossils until the Ordovician. And from that time to the present they have left only about 100 fossil species, so that there is nothing in their record to make us think that they were ever of much importance. The *scaphopods*, too, do not get well started until the Ordovician and have never amounted to much. The three major classes – gastropods, pelecypods, and cephalopods – all appear in the Cambrian. The first two groups increase steadily from that time to the present; and the GASTROPODS are probably very close to their peak of development now, comprising 49,000 of the 70,000 species of living molluscs. The CEPHALOPODS, though represented by only 400 living species, were once far more important than the gastropods; and they have left 10,000 fossil species to attest to their past glory. The dominant cephalopods of modern seas are squids and octopuses, which are probably more numerous at the present than they ever were. But it is the cephalopods with external shells, now represented by only 3 species of *Nautilus*, that made cephalopod history. The shells of the early *nautiloids* were straight cones internally divided into a series of chambers, as in the modern nautilus. In the Ordovician these shells were over 15 feet long, a length never again reached by any shelled invertebrate. Later the shells became slightly curved, then finally coiled. The nautiloids rose to a peak in the late Ordovician and Silurian and then declined; but they gave rise, it is believed, to another great group of coiled cephalopods, the *ammonoids*, which also went through an evolution from straight to coiled shells. The ammonoids became the dominant animals of Mesozoic seas and then died out in the Cretaceous. Some of the cretaceous ammonoids reverted to an uncoiled condition, so that the evolution of this group parallels that of gastropods (see p. 207). The advantage of coiling is probably the same in the two groups; it converts a long, unwieldly, straight cone into a compact, manageable coil.

EVEN the most advanced of invertebrates, the ARTHROPODS, must have had a long pre-Cambrian history, for at the beginning of the Cambrian we find three classes of arthropods already well started.

The most numerous of these are the TRILOBITES, which constitute over half of all Cambrian fossils. Apart from the undoubted success of these early arthropods, the abundance of their fossil remains is probably due in part to the fact that they moulted frequently and discarded numerous exoskeletons which were capable of fossilization. The name trilobite means 'three-lobed' and refers to the fact that the dorsal surface of the body is divided by two longitudinal furrows into three lobes. The body is also divided transversely into three regions: a head; a middle flexible portion, the thorax; and a posterior region, the abdomen, which consists of a number of fused segments and in some trilobites is prolonged into a spine. The head bears a pair of compound eyes, a pair of antennas, and four pairs of similar, jointed two-branched appendages. The outer branch, which is flattened and has a row of bristles along its posterior edge, is thought to have served for respiration and swimming. The inner branch was

probably used for walking. Similar appendages occur on all segments of the body, and they all have inwardly directed projections from the basal part of the limb. On the head appendages
these projections are modified for chewing food, as in modern
king crabs and arachnids. Most trilobites are from 1 to 3 inches
long, but some forms reached a length of over 2 feet. Their
habits can only be inferred; but it is thought that they lived in the
sea, since their remains are always found with corals, crinoids,
brachiopods, and other exclusively marine animals. Most trilobites probably inhabited shallow waters and were bottom-crawling types, either feeding on the various seaweeds, sponges,
coelenterates, brachiopods, and molluscs which are known to
have lived in the same places, or perhaps scavenging organic
debris by ploughing through the mud. The trilobites were the
dominant invertebrates of the Cambrian, and they continued to
flourish during the Ordovician, but then declined. They were
rare after the Devonian, and the last few survivors finally died
out in the Permian. It is perhaps no mere coincidence that their
decline followed the rise of the giant Ordovician cephalopods
and the hordes of Devonian fishes, both of which could have fed
on trilobites.

The trilobites probably gave rise to no other group of arthropods, but they seem most closely related to CRUSTACEANS. The
branchiopods, ancestors of the more primitive modern crustaceans like the fairy shrimps and cladocerans, are well represented
in lower Cambrian rocks. But the large crustaceans, such as lobsters and crabs, do not appear until the middle of the Mesozoic.

The AQUATIC ARACHNID-LIKE ARTHROPODS appeared in the
Cambrian; but their only living representative is *Limulus*, the
king crab, which has changed little since its appearance in the
Triassic, about 200,000,000 years ago. The most interesting of
these extinct arachnid-like animals are the EURYPTERIDS, some of
which attained a length of 10 feet, the largest size known for any
arthropod. It is not certain whether they were marine or inhabited
fresh-water streams, from where they were washed out to sea and
buried in marine sediments. They probably lived mostly on the
bottom, but their two large paddle-like appendages suggest that
they could swim. Eurypterids are known from the Cambrian,
flourished in the Silurian and Devonian, but, like the trilobites,
became extinct at the end of the Paleozoic.

The FIRST LAND ANIMALS may well have been certain primitive

air-breathing arachnids (Palaeophonus and Proscorpius from the Silurian), which resemble eurypterids in certain respects. The transition from book gills, like those of a limulus, to book lungs, like those of a scorpion, requires very few changes.

The fossil record of land arthropods, like that of other land animals, is a scanty one and does not necessarily give a true picture of either the numbers of species or the time of the earliest appearance of the different groups. Although about half a million species of living insects have been described, only a few thousand fossil insect species have been found. Most of these are from a few special regions where fossilization took place under very unusual conditions. The insect-containing amber of the Baltic region of Europe has already been mentioned; one of the best American localities for collecting insect fossils is at Florissant, Colorado, where falling volcanic ash from an ancient eruption carried land insects down into a lake and buried them in the mud at the bottom. In spite of their slim chances of fossilization, MILLIPEDES are known from the Devonian and CENTIPEDES and INSECTS from the Pennsylvanian. All of the early insects are now extinct, and only in a few cases can modern insects be assigned to an order which was already represented in the Paleozoic.

SINCE most ECHINODERMS have skeletons made of calcareous plates and live in shallow marine waters, the group has left one of the most informative of fossil records. The earliest echinoderms were, typically, sessile types which lived attached to the bottom, either directly or by a stem. The oldest and least specialized ones were the CYSTOIDS, which had an ovoid or globular shell composed of tightly fitting plates. On the surface of the shell were ciliated grooves (usually five) for food-collecting, and these extended on to the arms which were attached at the upper end of the shell. From the cystoids, which appear in the Cambrian and become abundant in the Silurian, are thought to have come the other two classes of stalked echinoderms, the BLASTOIDS and CRINOIDS. Not all stalked forms lived attached to the bottom; but the stalked, sessile types were the dominant echinoderms of Paleozoic times. The cystoids and blastoids became extinct by the end of the Paleozoic, and only the crinoids (sea lilies) have survived to the present. The earliest crinoids were all attached, with stems usually from 1 to 3 feet long, or over 70 feet long in at least one case. Fossil crinoids were known long before any living forms had been seen; and the class was believed to be extinct until, only a little over 60 years ago, a dredge brought up some living specimens. Although 90 per cent of modern forms are stemless and free-swimming, in all cases where the development has been studied there is an early stalked stage.

During the Paleozoic the free-living echinoderms – STARFISHES, SERPENT STARS, SEA URCHINS, and SEA CUCUMBERS – were inferior to the stalked types both in numbers of individuals and in species. But in the early Mesozoic they expanded rapidly and have maintained their superiority ever since.

To the geologist the fossil record serves not only as a means of determining the time of deposition of rocks in widely separated parts of the world, but also as a key to the study of ancient geography and climate. Fossil corals, echinoderms, brachiopods, and cephalopods always indicate the former presence of salt water. The occurrence of fossil coral reefs in Chicago is clear evidence that this region was once covered by a sea and that the climate at one time must have been much warmer than it is now. To biologists the fossil record furnishes abundant and direct evidence of the evolution of modern animals from simpler types which have preceded them. By examining one layer of rock after another, we can follow the early appearance of a group as a few simple, adaptable forms, which gradually increase in number, specialize, and radiate out into a variety of habitats, then finally degenerate into bizarre, over-specialized forms which, with the first radical change in the environment, die out altogether. Such is the evolutionary history, clearly recorded in the rocks, of the trilobites and ammonoids – animals which dominated the seas for millions of years, yet have left not a single descendant. There is no reason to doubt that many of the invertebrate groups flourishing at the present time are heading for the same fate.

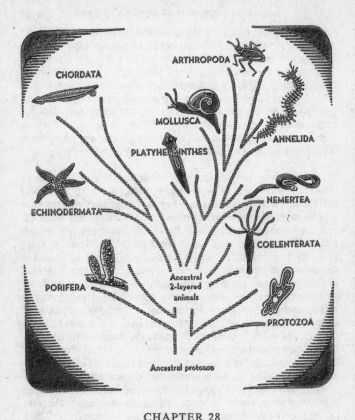

CHAPTER 28

Invertebrate Relationships

EVERYONE enjoys the unravelling of a good mystery, but no one would like to read on from clue to clue, until the earliest and most important events seemed about to be disclosed, only to find that the rest of the pages in the book were missing. Just this kind of exasperating situation confronts us when we try to relate animals to one another in an orderly scheme. Anyone can see that honeybees are much like bumble bees, that bees resemble flies more than they do spiders, and that spiders are more like lobsters than like clams. But when we attempt to relate the phyla, which, by definition, are groups of animals with fundamentally different body plans, there is little we can say with certainty. Arthropods are clearly allied to annelids, but how they are related to such utterly different animals as starfishes or vertebrates is quite obscure. The fossil record, which in many cases provides us with a whole series of gradually changing specimens from which we can work out the evolution of one small group from another, is of practically no use in relating the phyla to each other. For, as we dig deeper and deeper into the rocks, expecting to find a level at which the most recently evolved phyla no longer appear, we find instead that the fossil record is obliterated. The oldest rocks have stood the longest time and have been subjected to the greatest number of stresses, including those caused by the weight of the rocks above them; any fossils which they once contained have been changed beyond recognition. In the rocks of the earliest period for which we have good fossils (the Cambrian period), all the important invertebrate phyla are already represented. Thus, while the fossil record tells us a great deal about what the early representatives of the phyla were like, it has nothing to say about the order in which the phyla arose. Despite this, the situation is by no means hopeless. Good detectives have been known to reconstruct, in detail, the events leading up to a crime to which there were no witnesses. And biologists have been able to find some definite clues to events that happened considerably more than 500,000,000 years ago.

The most important kind of evidence is that based on a comparative study of the structure and development of the various groups. The use of such evidence is based on the assumption that the more closely the body plans of two phyla resemble each other, the closer their relationship and the more recent their common ancestor. This is the *principle of homology*, discussed in chapter 22. Sometimes the adult structure of two groups is so highly modified in adaptation to their different ways of life that the groups show very little similarity, and yet the early embryonic stages are almost identical. Here, also, we must assume a relationship, though a somewhat more remote one, for the early stages of development tend to be more conservative than the later ones, and a study of the embryology of animals often reveals basic similarities which otherwise would not have been suspected.

Related evidence is that based upon the PRINCIPLE OF RECAPITULATION, which states that 'ontogeny tends to recapitulate phylogeny.' Translating this into everyday English it reads: 'every animal, in its individual development from egg to adult, passes through a series of stages which correspond to stages in the long evolutionary history of its group.' This does not mean that the embryo of a man at any time resembles an adult fish or an adult reptile, but that it goes through stages which correspond to those undergone by the *embryos* of fishes and reptiles. The reason for this, apparently, is that the early developmental stages of an animal, in which the basic body structure is laid down, are less subject to modification than are the later stages in which the more superficial structures appear. Thus, the animals of a single phylum tend to look alike in their early embryonic stages and become gradually differentiated only later, as the structures peculiar to the different classes, and finally of the orders and still smaller categories, are produced. Moreover, even animals of different phyla appear so much alike in their very early development that they cannot be distinguished.

For example, a lobster starts out as a fertilized egg, a single cell which looks like the egg of any other invertebrate and is spherically symmetrical, showing no more differentiation than the simplest protozoan. The egg divides repeatedly, resulting in a blastula, a hollow ball, composed of a single layer of cells. We can find its counterpart among adult animals in the colonial protozoans like the volvox, a colony of flagellated cells arranged in a single layer on the surface of a hollow ball of jellylike material (pp. 50-1, Vol. I). The lobster blastula, by a proliferation of cells into the interior, soon becomes converted into a two-layer gastrula, much like that of any other animal. Since the gastrula has a depression at one end, which makes it radially symmetrical, and since it has two layers of cells, we may compare it with the coelenterate stage in evolution, though, of course, it has none of the specialized features of adult coelenterates. With the development of the mesoderm, and a differentiation of anterior and posterior ends, besides dorsal and ventral surfaces, the bilateral embryo has achieved the structural level of a flatworm. Next, segments appear, and pairs of similar appendages grow out, as in annelids. The appendages become two-branched; but, since they are mostly alike, the embryo reminds us of a primitive crustacean. With the differentiation and specialization of the appendages we finally recognize the developing animal to be a lobster. Thus, the development of a single individual is a condensed and modified recapitulation of what we believe was the evolutionary history of its phylum. In the same way we can hope to learn, from a study of the development of a planaria or a nereis, something of the evolutionary history of flatworms and annelids.

However, it should be pointed out at once that the PRINCIPLE OF RECAPITULATION HAS VERY DEFINITE LIMITATIONS. In the first place, no development occupying a few days or weeks could possibly go through every stage in an evolutionary history stretching over at least a milliard years. In the second place, not only adults but also embryos undergo evolution in adaptation to their environment. The mosquito larva, which lives in the water and feeds on debris, is a young stage modified for a specialized way of life that was never followed by any ancestor of the mosquito. Mosquitoes are descended from land insects, and the adults have had no connection with aquatic life since their primitive arthropod ancestors left the water. Moreover, many embryos develop special membranes and other structures which serve to protect or to nourish the young stages and have no counterpart in any adult animal. It has also been pointed out that, in the purely mechanical matter of changing from a single-celled zygote into a multicellular, many-layered animal of complex structure, all embryos must go through stages which are similar but do not necessarily indicate common descent for all the forms which exhibit them. Thus, in order to get a three-layered animal from a single cell, the cell has to form a one-layered organism, then a two-layered one, and finally a three-layered one; there is no other easy way to achieve such a result. And in the development from a spherical egg with all axes alike to a bilateral adult with three differentiated ones, there would have to be some intermediate radial stage with only one differentiated axis, even if the group had never had an ancestor which was radially symmetrical in the adult stage. Such arguments warn us that we must be cautious in assigning evolutionary significance to every step in embryonic development. Still the principle of recapitulation, when correctly interpreted, has explained the appearance and subsequent disappearance in the embryo of many seemingly useless structures, such as the tail or gill pouches of man. And it has contributed greatly to our understanding of the relationships of animals to one another.

BASING our ideas on the principles of homology and recapitulation, we are able to construct ANIMAL TREES or other schemes which attempt to show the order of evolution and the relationships between the phyla. Considering the remoteness of the events with which we are dealing, and the inconclusive nature of much of the evidence, it is clear that any 'invertebrate tree' must be considered highly speculative. The 'tree' presented at the beginning of this chapter is only one version; it will have to be changed in the future if new evidence turns up. At the same time, it ties together a great body of facts which would otherwise have less meaning and serves as a framework on which to hang what appears to be a fairly plausible account of the evolution of the main phyla of invertebrates.

It is highly probable that the capacity for photosynthesis was a characteristic of the ancestors of primitive organisms. From a hypothetical ancestral type of 'plant-animal', the exact nature of which is unknown, came at least two main lines of descent, the animal kingdom and the plant kingdom (except the simplest plants, such as the bacteria, as was discussed in chap. 1). The reason for this belief, as explained at the beginning of this book, is the similarity between primitive plants and primitive animals. By a LOSS OF CHLOROPHYLL (perhaps at several different times for different protozoan groups) and the development of a variety of locomotory and food-catching mechanisms, the animal kingdom arose. The most primitive animals are SINGLE CELLS, but we must remember that the modern *protozoa* have had a long evolutionary history and have undergone many changes before arriving at the condition in which we find them today.

The exact manner in which MULTICELLULARITY arose cannot now be determined. But it is easy to understand how it could have evolved through the failure of individual cells to separate completely after division. Such colonies of attached but relatively independent cells are known to occur among the protozoa (see colonial collar flagellates on pp. 53-4, Vol. 1). In the volvox colony, already mentioned, there is a certain amount of co-operation in locomotion and in reproduction, but not enough to elevate the colony to the ranks of multicellular organisms. The volvox colony is highly specialized and must not be thought of as ancestral in any sense, but it gives us some idea of what one type of primitive multicellular organism might have been like.

In the *sponges*, the least integrated of the truly many-celled animals, the cells show considerable division of labour, but tissues are poorly developed, and the animals cannot be said to have gone much beyond a CELLULAR LEVEL OF ORGANIZATION. The porous construction, peculiar method of feeding, and the lack of a definite mouth and digestive cavity are among the reasons for thinking that sponges have no direct relationship to other animals. Perhaps they evolved from primitive collared flagellates, whose modern representatives are the only animals besides sponges which have collar cells.

Since sponges are not on the main line of evolutionary advance, the stage beyond the first multicellular organisms which led to the higher phyla can only be imagined. By the passage of some of the cells from the surface into the interior, a TWO-LAYERED ANIMAL was formed. This hypothetical two-layered ancestor probably evolved from a different group of protozoa from that which gave rise to sponges. Constructed on the TISSUE LEVEL OF ORGANIZATION, with an outer ciliated ectoderm specialized for locomotion, protection, and sensation, and an inner endoderm specialized for digestion, it was master of the ancient seas as it swam about, feeding on protozoans and unicellular plants. Just what it looked like we do not know, but it probably resembled the radially symmetrical, ciliated, two-layered, free-swimming larva of the obelia and most other marine *coelenterates*. The wide occurrence of such a larva in the embryology of coelenterates indicates that this phylum probably arose from a simple two-layered ancestor by the outgrowth of tentacles round the mouth.

The next stage in evolutionary advance seems to have been the formation of the MESODERM, a tissue between the ectoderm and endoderm, from which more definite organs and ORGAN-SYSTEMS could be made, resulting in greater size and complexity of construction. Mesoderm develops in animals by one of two main ways. In the flatworms, nemerteans, molluscs, annelids, and arthropods it usually originates from two special cells, known as 'primitive mesoderm cells', set aside in the early gastrula. In the Echinodermata, the phylum to which the starfish belongs, and in the Chordata, the phylum to which man belongs, the mesoderm comes from outpocketings of the primitive endoderm. Because of this difference in mesoderm formation, among other reasons, we recognize two great lines of evolution known as the *arthropod* and the *chordate lines*, and indicated on the animal tree by a main branching.

Following the ARTHROPOD LINE of evolution, we see the *flatworms* exhibiting the beginning of the importance of the mesoderm. Here the first organ systems are differentiated, and accompanying this increased complexity, is *bilateral symmetry*. It is not known just how the radial gastrula-like ancestor became bilateral, but specialization of one end and a bottom-crawling habit were probably stages in the process.

We come next to the *nemerteans*. From many details of adult structure and early development, nemerteans can be regarded as closely related to flatworms, but they show two distinct advances:

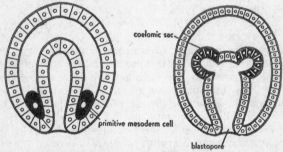

coelomic sac

primitive mesoderm cell

blastopore

ORIGIN OF THE MESODERM in arthropod and chordate lines. *Left*, the embryo of a mollusc showing the primitive mesoderm cells budded off from the primitive endoderm. *Right*, an echinoderm larva showing the mesoderm being budded off as pouches from the sides of the primitive endoderm.

the development of a digestive tract with two openings, and the beginning of a circulatory system. Much more advanced are the *molluscs* (snails, slugs, clams, oysters, squids, octopuses, etc.), which have further specialized and perfected the systems pioneered by nemerteans. Molluscs are not segmented and therefore must have diverged from the main line of evolution before SEGMENTATION arose. Thus, on the animal tree they are shown to branch off at a point below that from which *annelids* diverge. In spite of the great lack of structural similarity, molluscs are closely related to annelids – one of the few relationships between any two invertebrate phyla for which we have indisputable evidence. The early embryos of annelids and molluscs are almost identical, cell for cell. The mesoderm arises from a corresponding

cell in both groups; and their free-swimming larvas, the TRO-
CHOPHORES, are very much alike (chap. 19). A trochophore-like
larva, with a ciliated band around the equator, is characteristic
not only of annelids and molluscs but also of a number of minor
phyla (discussed in chap. 16) not shown on the 'tree'. And the
larvas of flatworms and nemerteans are certainly much like the
trochophore. Thus, the trochophore type of larva serves to link
together a whole series of phyla.

The *arthropods* have no trochophore, and they show few
similarities to annelids in the early stages of development; but
their adult structure is similar in so many respects that there can
be no doubt that the two groups had a common segmented
ancestor with a pair of appendages to each segment and a nervous system which encircled the anterior end of the digestive
tract and passed backward along the ventral surface as a double
cord with segmental ganglia.

The CHORDATE LINE includes only two main phyla: the echinoderms and chordates. This means that the invertebrate phylum
most closely related to man and the vertebrates is one which includes such animals as the starfishes. The reasoning by which we
arrive at this rather surprising conclusion is based mainly on a
comparison of the developing embryos of certain members of the
two groups. In the first place, as already mentioned, the mesoderm of echinoderms and chordates arises in the same way. Also,
the COELOM is formed in both groups from the hollow meso-
dermal pouches, whereas in the arthropod line it arises by splits
in the bands of mesoderm budded off from the primitive mesoderm cells. In either case the end result is the same, and, by in-
specting the adult animals, it cannot be told that their coeloms
arise in different ways. In addition, the free-swimming larva of
echinoderms is quite different from the trochophore larva
characteristic of annelids, molluscs, and some of the other phyla
of the arthropod line. The bilateral echinoderm larva is more
flattened than the trochophore and has longitudinal, looped
ciliated bands for locomotion. Since this same type of larva
occurs in all classes of echinoderms, it is believed to resemble a
hypothetical ancestral type (dipleurula) from which all modern
echinoderms have been derived.

The phylum *Chordata* consists mainly of vertebrates but includes three groups of invertebrates: acorn worms, tunicates,
and the amphioxus. These groups have no backbone but are

classed with the vertebrates because, as described in chapter 26, they possess, at some time in their life-history, a stiffening rod (the notochord), pharyngeal gill pouches or slits, and a dorsal tubular nerve cord. Now, it must be clearly understood that echinoderms have none of these structures; their affinity with chordates is based on similarities in the development of the mesoderm and coelom mentioned before and on the slender clue afforded by a striking resemblance between the (dipleurula) type of larva found in echinoderms and the larva of the acorn worm. As was pointed out in the discussion of recapitulation, larval resemblances may be misleading, because larvas themselves

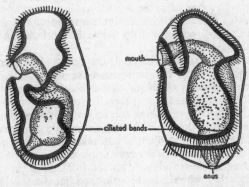

mouth

ciliated bands

anus

LARVA OF AN ECHINODERM.   LARVA OF AN ACORN WORM.
(After Bury.)   (After Morgan.)

undergo evolution in adaptation to their environment. Since the larvas of echinoderms and acorn worms both live in the surface waters of the ocean, feeding on microscopic organisms, their similarities may be independent responses to the same conditions of life. On the other hand, the larvas of flatworms, nemerteans, molluscs, and annelids have for over 500 million years lived in the same places and fed in the same way; yet all conform to the trochophore type. This leads us to believe that the differences between these two larval types have real evolutionary significance and are the result of a divergence of two main stocks from some ancestral bilateral animal in the remote past.

Since the echinoderms are not clearly segmented, the segmentation of chordates must have arisen after the two groups which finally gave rise to the modern echinoderms and chordates had already separated. The segmentation of chordates results from the budding of a series of mesodermal pouches from the primitive endoderm. In annelids and arthropods it forms by a crosswise breaking-up of continuous bands of mesoderm. For this reason, and the fact that the chordate and arthropod lines are thought to have separated from each other long before segmentation arose, segmentation of the chordates and of the annelids and arthropods is not considered homologous but must have evolved independently in the two groups.

In the earliest schematizations of animal evolution, the animals were usually pictured as ascending, on a vertical ladder, directly 'from amoeba to man', with the other animals placed on intermediate rungs according to their order of increasing complexity. Now it is realized that animals do not form a continuous series and that their relationships to one another are more correctly represented by a branching 'tree'. The ancestral groups, long extinct, are placed on the main stems. Extinct forms now represented only by fossils are placed on dead side branches (not shown in the 'tree' given here). On the ends of the living side branches are placed the modern forms. A 'tree' not only fits the facts more closely than a ladder, but its many separate branches, representing independent lines of specialization, help to suggest why we find such endless variety among the animals without backbones.

INDEX

Numbers in italics refer to the photographic insets. Generic and specific names of animals are printed in italics.

MORE ABOUT PENGUINS
AND PELICANS

Penguinews, which appears every month, contains details of all the new books issued by Penguins as they are published. From time to time it is supplemented by *Penguins in Print*, which is a complete list of all books published by Penguins which are in print. (There are well over three thousand of these.)

A specimen copy of *Penguinews* will be sent to you free on request, and you can become a subscriber for the price of the postage. For a year's issues (including the complete lists) please send 25p if you live in the United Kingdom, or 50p if you live elsewhere. Just write to Dept EP, Penguin Books Ltd, Harmondsworth, Middlesex, enclosing a cheque or postal order, and your name will be added to the mailing list.

A Penguin Reference book is described overleaf.

Note: *Penguinews* and *Penguins in Print* are not available in the U.S.A. or Canada

A DICTIONARY OF BIOLOGY

M. Abercrombie, C. J. Hickman, and M. L. Johnson

5TH REVISED EDITION

In this dictionary the author's aim is to explain biological terms which a layman may meet when reading scientific literature, and to define the terms which a student of biology has to master at the beginning of his career – the thousand or so words which so grimly guard the approaches to the science. The entries are not restricted to a bare definition, which, however valuable in keeping one to the narrow path of customary usage, can be most unhelpful in the task of understanding the word in its context. Some information about most of the things named is given, so as to convey something of their significance in biological discussion – to add a faint flavour of a pocket encyclopedia.

The most important thing the authors have not tried to do is to lay down the law. They have tried to interpret a foreign language as it is actually used. It would be valueless and presumptuous to define the words according to how they think they ought to be used. It would be wrong also to rely on etymology as a guide to correct usage. The meaning of a Greek root may be unequivocal, but biologists are not talking Greek; they are using a living language, and the proof of the meaning is in the speaking. The authors admit that the terminology is sometimes a little messy (e.g. Lipoid) and that there is often a disconcerting disparity between botany and zoology (e.g. Cell); but they have recorded what seem to them to be the customary use or uses of the term at the present day.